13th International Conference on Quasicrystals (ICQ13)

Journal of Physics: Conference Series Volume 809

Kathmandu, Nepal
18 – 23 September 2016

Editors:

An Pang Tsai
Hem Raj Sharma
Ron Lifshitz
Ryuji Tamura

ISBN: 978-1-5108-3856-7
ISSN: 1742-6588

Printed from e-media with permission by:

Curran Associates, Inc.
57 Morehouse Lane
Red Hook, NY 12571

CURRAN ASSOCIATES INC. proceedings.com

Some format issues inherent in the e-media version may also appear in this print version.

Printed by Curran Associates, Inc. (2017)

Institute of Physics
Dirac House, Temple Back
Bristol BS1 6BE UK

Phone: 44 1 17 929 7481
Fax: 44 1 17 920 0979

techtracking@iop.org

Additional copies of this publication are available from:

Curran Associates, Inc.
57 Morehouse Lane
Red Hook, NY 12571 USA
Phone: 845-758-0400
Fax: 845-758-2633
Email: curran@proceedings.com
Web: www.proceedings.com

Table of contents

Volume 809

13th International Conference on Quasicrystals
18–23 September 2016, Kathmandu, Nepal

Accepted papers received: 7 February 2017
Published online: 28 March 2017

Preface

011001
OPEN ACCESS
13th International Conference on Quasicrystals

An Pang Tsai, Hem Raj Sharma, Ron Lifshitz and Ryuji Tamura

011002
OPEN ACCESS
Peer review statement

Papers

012001
OPEN ACCESS
Quasicrystal structure and growth models: discussion of the status quo and the still open questions

Walter Steurer.....1

012002
OPEN ACCESS
Cluster environments in a twelve-fold packing model of icosahedral quasicrystals

Hiroyuki Takakura and Radosław Strzałka.....7

012003
OPEN ACCESS
Origin of 18-fold quasicrystal

S Bekku, P Ziherl and T Dotera.....11

012004
OPEN ACCESS
Crystal structure of χ-AlNiRu containing pseudo-Mackay Clusters

S Suzuki, R Simura, A Yoshiasa and K Sugiyama.....18

012005
OPEN ACCESS
Medium Range Ordering realized in $Zr_{80}Pt_{20}$Amorphous Alloy

K Sugiyama, T Kawamata and T Muto.....22

012006
OPEN ACCESS
Foundations of the quasicrystalline structure description in the statistical approach

Janusz Wolny, Ireneusz Buganski, Maciej Chodyn and Radoslaw Strzalka.....26

012007
OPEN ACCESS
Atomic structure of the primitive cubic phase P_{40} in the Al-Pd-Ru system

Y Hatakeyama, N Fujita and A P Tsai.....33

012008
OPEN ACCESS
Contribution to thermodynamic description of Al-Pd system

Libor Ďuriška, Ivona Černičková, Roman Čička and Jozef Janovec.....37

012009
OPEN ACCESS
Experimental study of phases I and H in selected (56-60)Zn-(29-35)Mg-(8-13)Y alloys after long-term annealing at 400°C

Marek Adamech, Ivona Černičková, L'ubomír Čaplovič, Pavol Priputen, Dusan Janičkovič and Jozef Janovec.....41

012010
OPEN ACCESS
A new superlattice structure in the $Al_{2.75}Ir$ and $Al_{2.63}Rh$ 1/0 approximants

Kazue Nishimoto, Natsumi Yamaguchi and Ryuji Tamura.....45

012011
OPEN ACCESS
Synthesis characterization and hydrogenation behaviour of as quenched $Ti_{41.5+X}Zr_{41.5-X}Ni_{17}$ (x=0, 3.5, 11.5 and 13.5) nano quasicrystalline ribbons

Rohit R Shahi, T.P. Yadav, M.A. Shaz and O.N. Srivastava.....49

012012
OPEN ACCESS
Magnetic-field dependence of Schottky specific heat in the 1/1 Zn-Sc-Tm approximant

J Dolinšek.....53

012013
OPEN ACCESS
Magnetic Properties of the Au-Al-Yb Approximant under Hydrostatic Pressure

Shuya Matsukawa, Kazuhiko Deguchi, Keiichiro Imura, Tsutomu Ishimasa and Noriaki K. Sato.....57

012014
OPEN ACCESS
Unified cluster-based description of valence bands in AlIr, $RuAl_2$, $RuGa_3$ and Al–TM quasicrystalline approximants

K Kitahara, Y Takagiwa and K Kimura.....62

012015
OPEN ACCESS
Aging effects of dodecagonal quasicrystal formed in Mn-Cr-Ni-Si alloys

Kotoba Toyonaga and Tsutomu Ishimasa.....66

012016
OPEN ACCESS
Preparation dependent surface structure of NiAl(100)

Sam Coates, Abdullah Al-Mahboob, Ronan McGrath and Hem Raj Sharma.....70

012017
OPEN ACCESS
Medium energy ion scattering (MEIS) study from the five-fold surface of icosahedral Ag-In-Yb quasicrystal

H. R. Sharma, P. J. Nugent, T. C. Q. Noakes, J. Smerdon, J. Parle, A. P. Tsai and R. McGrath.....75

012018
OPEN ACCESS
Adsorption structure of Bi on the fivefold surface of i-Ag-In-Yb quasicrystal

Kazuki Nozawa and Yasushi Ishii.....79

012019
OPEN ACCESS
Phason dynamics in decagonal quasicrystals

Hans-Rainer Trebin and Hansjörg Lipp.....84

012020
OPEN ACCESS
Sub-diffusive electronic states in octagonal tiling

G. Trambly de Laissardière, C. Oguey and D. Mayou.....88

012021
OPEN ACCESS
Screening of a magnetic impurity in a quasicrystal

Anuradha Jagannathan.....92

012022
OPEN ACCESS
Cluster Size Effects on Electronic Reconstruction in Quasiperiodic System

Ryu Shinzaki, Joji Nasu and Akihisa Koga.....99

012023
OPEN ACCESS
Gap structure of 1D cut and project Hamiltonians

Nicolas Macé, Anuradha Jagannathan and Frédéric Piéchon.....103

012024
OPEN ACCESS
Brick tiling

S I Ben-Abraham and D Flom.....107

012025
OPEN ACCESS
Color coding for the brick tiling

D Flom and S I Ben-Abraham.....111

012026
OPEN ACCESS
Diffraction of a binary non-Pisot inflation tiling

Michael Baake and Uwe Grimm.....115

012027
OPEN ACCESS
Substitution-based sequences with absolutely continuous diffraction

Lax Chan and Uwe Grimm.....119

012028
OPEN ACCESS
An ancient rule for constructing dodecagonal quasiperiodic patterns

Rima Ajlouni.....123

012029
OPEN ACCESS
Color groups of colorings of *N*-planar modules

Manuel Joseph C. Loquias, Lilibeth D. Valdez and Ma. Lailani B. Walo.....128

012030
OPEN ACCESS
Electron Localization and Conduction Mechanism of Icosahedral Quasicrystal Al-Pd-Ru

Y. Sekiyama and I. Kanazawa.....132

IOP Conf. Series: Journal of Physics: Conf. Series **809** (2017) 011001 doi:10.1088/1742-6596/809/1/011001

Journal of Physics: Conference Series

The 13th International Conference on Quasicrystals (ICQ13)

18-23 September 2016

Kathmandu, Nepal

Editors

An Pang Tsai

Tohoku University, Sendai, Japan

Hem Raj Sharma

The University of Liverpool, Liverpool, UK

Ron Lifshitz

Tel Aviv University, Israel

Ryuji Tamura

Tokyo University of Science, Tokyo, Japan

IOP Conf. Series: Journal of Physics: Conf. Series **809** (2017) 011001 doi:10.1088/1742-6596/809/1/011001

Preface

The 13th International Conference on Quasicrystals (ICQ13) was held in Kathmandu, Nepal, from 18th to 23rd September 2016. This episode of a series of conferences on quasicrystals followed twelve previous successful meetings. The first event, the "International Workshop on Aperiodic Crystals", was held in Les Houches, France in 1986, which was followed by the "International Workshop on Quasicrystals" in Beijing, China in 1987 and the "International Meeting on Quasicrystals" in Vista-Hermosa, Mexico in 1989. Thereafter, a series of "International Conference on Quasicrystals" took place respectively in St. Louis, Missouri, USA in 1992 (ICQ4), Avignon, France in 1995 (ICQ5), Tokyo, Japan in 1997 (ICQ6), Stuttgart, Germany in 1999 (ICQ7), Bangalore, India in 2002 (ICQ8), Ames, Iowa, USA in 2005 (ICQ9), Zürich, Switzerland in 2008 (ICQ10), Sapporo, Japan in 2011 (ICQ11) and Cracow, Poland in 2013 (ICQ12).

The conference covered a wide range of topics on quasicrystals: formation, growth, and phase stability; structure and modeling; mathematics of quasiperiodic and aperiodic structures; transport, mechanical and magnetic properties; and finally, surfaces and overlayer structures. In addition, other aspects of material science such as metamaterials, polymer science, metallic glasses, and metallic alloys were discussed. The conference also reviewed the state-of -art of quasicrystal research and its future direction.

ICQ13 was attended by 98 researchers from 22 countries, including 26 PhD or master students. There were 13 invited talks (including a public lecture by the 2011 Nobel

Laureate in Chemistry, Prof. Dan Shechtman), 3 tutorial lectures, 50 contributed talks and 36 poster presentations. The conference was truly interdisciplinary - comprising of theoretical and experimental physicists, chemists, material scientists, and mathematicians.

The proceedings of the conference are published as open access in the Institute of Physics Conference Series Volume 809 (2017). The papers have been reviewed in the regular refereeing process. Among 35 proceedings received, 30 have been accepted for the publication.

The conference itself was held at the Dhulikhel Lodge Resort. The public lecture was organised at the central campus of Trubhuvan University in Kirtipur, hosted jointly by the Nepal Physical Society and the Central Department of Physics, while the tutorial lectures were held at Kathmandu University. These two events were open to local researchers and students. A special ceremony was organised for the inauguration of ICQ13, which was addressed and attended by the vice-chancellors of Tribhuvan University, Kathmandu University, Nepal Academy of Science and Technology (NAST), and representatives of various professional organisations, including the Nepal Government.

The Jean-Marie Dubois Award for Excellence in Quasicrystals (2016) was presented to Dr. Marek Mihalkovič at ICQ13, for the theoretical work that has enabled and demonstrated the simulation of thermodynamic and dynamic properties of quasicrystals - based on realistic atomic-scale models and energetics. The Jean-Marie Dubois Award was established to recognize important, sustained research on any aspect of quasicrystals in the last 10-year period preceding the award. The prizes for best student presentation were awarded to Ito Natsu from Toyo University of Science, Japan for her paper titled ''Synthesis and magnetic properties of the Au-Al-Yb approximant'' and Nicolas Macé from Université Paris-Saclay, France for his paper titled ''Gap structure and topological indices on the Fibonacci quasicrystal''.

We gratefully acknowledge the generous financial support for ICQ13 from the International Union of Crystallography (IUCr), the European Integrated Center for the

Development of New Metallic Alloys and Compounds (C-MAC), the International Centre for Diffraction Data (ICDD), Rigaku Oxford Diffraction, Embassy of Israel Kathmandu, Deutsch-Nepalische Gesellschaft, TBI Group, Phul Kumari Mahato Memorial Trust, Nepal Science Foundation Trust, Holy-Cow, officials of Non-Resident Nepali Association (NRNA) and NRNA-Australia.

We would like to express our sincere thanks to the members of the Organising Committee, Program Committee and International Advisory Board, invited speakers, session chairs, editorial board members and reviewers of the conference proceedings and all others for their support and involvement in the conference.

Last, but not least, we would like to thank the Local Organisation Committee of ICQ13 - chaired by Prof. Narayan Prasad Adhikari - for the efforts that made the conference so successful. We would also like to express our gratitude for the support we received from the University of Liverpool, UK, Tohoku University, Japan and Tribhuvan University, Nepal.

We hope that the conference helped to promote the struggling tourism sector of Nepal, after the country was hit by devastating earthquakes in 2015. We thank the Nepal Tourism Board, Atithi Devo Bhava, Chandragiri Cable Car and Cultural Bands to help ICQ13 participants explore the culture, history and natural beauty of the Himalayan country Nepal during their stay.

The 14th International Conference on Quasicrystals (ICQ14) will be held on 26 – 31 May 2019 in Bled, Slovenia.

Thank you.

Conference Chairs,
Dr Hem Raj Sharma (the University of Liverpool, UK)
Prof An Pang Tsai, Tohoku University, Japan
31st January 2017

IOP Conf. Series: Journal of Physics: Conf. Series **809** (2017) 011001 doi:10.1088/1742-6596/809/1/011001

Committees

Conference Chairs

Dr. Hem Raj Sharma
(The University of Liverpool, UK)
Prof. An Pang Tsai
(Tohoku University, Japan)

Conference Secretary

Ms. Noriko Yubuta
(Tohoku University, Japan)

International Advisory Board

Michael Baake (Germany)
Marc de Boissieu (France)
Janez Dolinšek (Slovenia)
Chuang Dong (China)
Jean-Marie Dubois (France)
Michael Feuerbacher (Germany)
Uwe Grimm (UK)
Yasushi Ishii (Japan)
Tsutomu Ishimasa (Japan)
Marek Mihalkovic (Slovakia)
Ron Lifshitz (Israel)
Ronan McGrath (UK)
N. K. Mukhopadhyay (India)
Walter Steurer (Switzerland)
Patricia Thiel (USA)
An Pang Tsai (Japan)
Michael Widom (USA)
Janusz Wolny (Poland)

Program Committee

Marc de Boissieu (France)
Vincent Fournee (France)
Hiroyuki Takakura (Japan)
Ron Lifshitz (Israel)
Hem Raj Sharma (UK)
An Pang Tsai (Japan)

Local Organising Committee

Chair: Prof. Narayan Prasad Adhikari (Central Department of Physics, TU)

Members:

Prof. Amar Yadav (Central Department of Physics, TU)
Dr. Rajendra Parajuli (Amrit Science Campus)
Dr. Gopi Chandra Kafle (Central Department of Physics, TU)
Dr. Lila Pradhan Joshi (Amrit Science Campus)
Dr. Raju Adhikari (Nepal Science Foundation (NSF) and CSIRO Manufacturing, Australia)

IOP Conf. Series: Journal of Physics: Conf. Series **809** (2017) 011001 doi:10.1088/1742-6596/809/1/011001

Conference Photo

Contents

First Author Surname	Paper Title	Page no
Steurer	Quasicrystal structure and growth models: discussion of the status quo and the still open questions	12001
Takakura	Cluster environments in a twelve-fold packing model of icosahedral quasicrystals	12002
Bekku	Origin of 18-fold quasicrystal	12003
Suzuki	Crystal structure of χ-AlNiRu containing pseudo-Mackay Clusters	12004
Sugiyama	Medium Range Ordering realized in $Zr_{80}Pt_{20}$ Amorphous Alloy	12005
Wolny	Foundations of the quasicrystalline structure description in the statistical approach	12006
Hatakeyama	Atomic structure of the primitive cubic phase P_{40} in the Al-Pd-Ru system	12007
Ďuriška	Contribution to thermodynamic description of Al-Pd system	12008
Adamech	Experimental study of phases I and H in selected (56-60)Zn-(29-35)Mg-(8-13)Y alloys after long-term annealing at 400°C	12009
Toyonaga	Aging effects of dodecagonal quasicrystal formed in Mn-Cr-Ni-Si alloys	12010
Nishimoto	A new superlattice structure in the $Al_{2.75}Ir$ and $Al_{2.63}Rh$ 1/0 approximants	12011
Shahi	Synthesis characterization and hydrogenation behaviour of as quenched $Ti_{41.5+X}Zr_{41.5-X}Ni_{17}$ (x=0, 3.5, 11.5 and 13.5) nano quasicrystalline ribbons	12012
Dolinšek	Magnetic-field dependence of Schottky specific heat in the 1/1 Zn-Sc-Tm approximant	12013
Matsukawa	Magnetic properties of the Au-Al-Yb approximant under hydrostatic pressure	12014
Kitahara	Unified cluster-based description of valence bands in AlIr, $RuAl_2$,	12015

IOP Conf. Series: Journal of Physics: Conf. Series **809** (2017) 011001 doi:10.1088/1742-6596/809/1/011001

	$RuGa_3$ and AlTM quasicrystalline	
Coates	Preparation dependent surface structure of NiAl(100)	12016
Sharma	Medium energy ion scattering (MEIS) study from the five-fold surface of icosahedral Ag-In-Yb quasicrystal	12017
Nozawa	Adsorption structure of Bi on the fivefold surface of i-Ag-In-Yb quasicrystal	12018
Trebin	Phason dynamics in decagonal quasicrystals	12019
Laissardière	Sub-diffusive electronic states in octagonal tiling	12020
Jagannathan	Screening of a magnetic impurity in a quasicrystal	12021
Shinzaki	Cluster Size Effects on Electronic Reconstruction in Quasiperiodic System	12022
Macé	Gap structure of 1D cut and project Hamiltonians.	12023
Ben-Abraham	Brick tiling	12024
Flom	Color coding for the brick tiling	12025
Baake	Diffraction of a binary non-Pisot inflation tiling	12026
Chan	Substitution-based sequences with absolutely continuous diffraction	12027
Ajlouni	An ancient rule for constructing dodecagonal quasiperiodic patterns	12028
Loquias	Color groups of colorings of N-planar modules	12029
Sekiyama	Electron Localization and Conduction Mechanism of Icosahedral Quasicrystal Al-Pd-Ru	12030

Peer review statement

All papers published in this volume of *Journal of Physics: Conference Series* have been peer reviewed through processes administered by the proceedings Editors. Reviews were conducted by expert referees to the professional and scientific standards expected of a proceedings journal published by IOP Publishing.

Quasicrystal structure and growth models: discussion of the status quo and the still open questions

Walter Steurer
Department of Materials, ETH Zurich, Leopold-Ruzicka-Weg 4, 8093 Zurich, Switzerland
E-mail: steurer@mat.ethz.ch

Abstract. Where are we now in quasicrystal (QC) research more than three decades after Dan Shechtman's discovery? Do we fully understand the origin of quasiperiodicity, the formation, growth, thermodynamic stability, structure and properties of quasicrystals? First, I will shortly present the status quo, then I will address the still open questions, and identify potential focus areas for future research. Because of the limited space, I will focus on decagonal quasicrystals (DQCs); the status quo for research on icosahedral quasicrystals (IQCs) is comparable.

1. Introduction

More than three decades after Dan Schechtman's discovery [1], QC research has become a mature field. This is reflected in the approximately 12,000 publications that have appeared so far (for an introduction into the area and an overiew of the field see [2], for instance). That the number of QC publications per year is shrinking since the peak year 2000 (Fig. 1) does not mean that we know now more or less everything what we want to know about QCs; it is rather a consequence of the lack of important applications for intermetallic QCs and related therewith the ever decreasing funding.

While we already know many fascinating details about the yet increasing number of QC, the big picture is still showing many white spots. Even some fundamental questions are still open:

- What is the origin of quasiperiodic order?
- How is quasiperiodicity achieved during growth?
- What is the role of entropy for the stability of QCs?
- Are there many more yet unknown intermetallic systems featuring QCs?
- Why is fivefold symmetry prevalent in intermetallic QCs in contrast to twelvefold symmetry for assemblies on the meso- or macro-scale?
- Etc.

In contrast to intermetallic QCs, quasiperiodic arrangements on the mesoscale are much better understood due to the simple, well-defined pair-potentials between the interacting particles, which can be easily modeled (see, e. g., [3] and references therein). The atomic interactions in intermetallic QCs are much more complex and diverse, and they can vary significantly in the structure between covalent and metallic bonding. Furthermore, the global valence electron concentration (VEC) plays a decisive role for the stability of intermetallic QCs [4]. In the following, I will address the topics 'QC structure analysis', 'QC stability' and 'QC growth', and suggest ways for tackling the open questions.

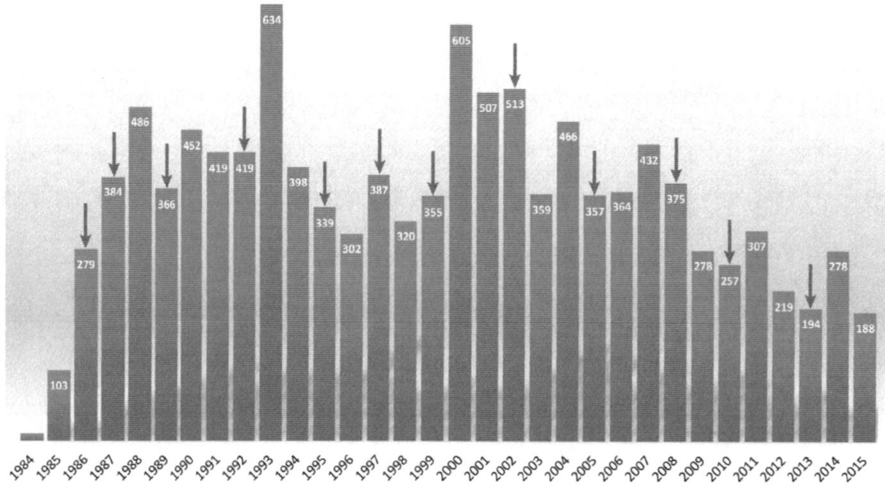

Figure 1. Number of publications on QCs per year. The arrows mark the years when International Conferences on Quasicrystals (ICQs) took place. Proceedings of these conferences appeared in most cases in the subsequent year increasing the number of papers significantly over the average value in several cases.

2. QC structure analysis

State-of-the-art structure analysis of complex intermetallic phases in general does not only comprise the quantitative determination of the atomic positions and displacement parameters but also of the chemical bonding and the electronic structure (see, e.g., [5] and references therein). Furthermore, for a full understanding of the structure also the kind of intrinsic disorder, if any, has to be determined. The method of choice for QC structure analysis is based on X-ray and/or neutron diffraction in combination with spherical-aberration (C_s) corrected electron microscopy. Needless to say that only thermally well-equilibrated homogeneous samples should be used. This method allows obtaining quantitative information on the local as well as the global arrangement of atoms and structural subunits (clusters) in a quasiperiodic structure. However, one has to be aware of the limits of this approach:

Diffraction techniques

(i) Once one identifies the observed reflection intensities as Bragg reflections, one excludes *per definitionem* all structure models that can only be described as decorated non-Pisot (random) tilings (see, e.g., [6]).

(ii) Applying the higher-dimensional (nD) approach automatically leads to quasiperiodic structure models.

(iii) If only Bragg reflection data are used, then the structural information is averaged over the irradiated volume of the QC modulo the nD unit cell. Only electron microscopy can show the differences between the (over the sample thickness projected) actual structure and the so obtained averaged quasiperiodic structure (for an example see, e.g., [7]).

(iv) Due to the scaling symmetry of the diffraction pattern, the choice of a proper unit cell leading to the optimum partition (large contiguous areas) of the atomic surfaces (occupation domains) is non-trivial. This problem can be solved based on the par(allel)-space pair-correlation (Patterson) function, which allows to derive the typical intercluster distances [8, 9].

(v) If the cluster-embedding method is used, one has to keep in mind that the clusters identified in approximants (ACs) do not necessarily correspond one-to-one to the clusters in the related QC (see also [10, 11]). For instance, in the case of decagonal Al-Co-Ni and Al-Co-Cu the cluster diameters of the ACs amount to ≈ 20 Å diameter (the size is already limited by the unit cell dimensions) [12], while those of the DQCs have been found to be τ times larger [13, 15].

(vi) Dynamical effects such as multiple diffraction can change the intensity of weak reflections drastically (see, e.g., [16] and references therein). If not taken into account it can deteriorate the quality of a refined structure model significantly.

(vii) Using improper weighting schemes for weak reflections can lead to too large phason Debye-Waller factors masking failures in the refined structure models. Reflections with large perp(endicular)-space components of the diffraction vectors have low intensities. If they are biased, it will have implications for the quality of the partitioning of the atomic surfaces, and therewith also for the determination of the kind of quasiperiodic long-range order.

Electron microscopy

(i) The structural information is averaged over the whole sample thickness (≈ 10 nm corresponding to 40-50 atomic layers). The imaged area may not be characteristic for the whole sample.

(ii) Whole columns of atoms (parallel to the electron beam) must flip at the same time to make a flip observable. Such events will be much rarer than single atom flips.

(iii) Contrasts due to structural disorder between different atomic layers perpendicular to the electron beam will be leveled down due to averaging and may become unobservable.

Open questions and ways to answer them

QC structure analysis should focus on the determination of the actual long-range order and its variation as a function of temperature. So far, it seems to have been implicitly assumed that the 2D Penrose tiling and the 3D Ammann tiling, respectively, are good model quasilattices for the actual structures of most QCs. Only in a few cases other kinds of quasiperiodic order have been explicitly suggested (see, e.g., [13]). However, while the long-range order of periodic intermetallics is already defined by the determination of the Bravais lattice, one out of 14, this is more complex for QCs. In the case of DQCs, for instance, there exists an infinite number of different quasiperiodic tilings that can underly the structure; for instance, the generalized Penrose tilings [17] or the Masakova tilings [18, 19, 20].

It is also an open question whether structural disorder mainly takes place between atoms within clusters or between clusters. Can structural disorder destroy clusters or are clusters relatively stable subunits? How does the disorder change as a function of temperature? Favorable ways to determine the actual, non averaged long-range order are, for instance:

(i) Large-scale electron microscopic studies for the determination of the cluster ordering (see, e. g., [14]).

(ii) Large-scale surface studies by scanning tunneling microscopy (STM) or atomic force microscopy (AFM) for the determination of the cluster ordering.

(iii) Repeated scans of one and the same area of a QC sample by STM/AFM at elevated temperature for identifying any changes in the quasiperiodic order (rearrangements of clusters). In contrast to electron microscopy, where whole columns of atoms must flip to be observable, single-atom flips can be detected by STM.

3. QC stability

It has been a long-lasting discussion whether QCs are energy- or entropy-stabilized, i.e., whether or not the entropic contribution is essential for their thermodynamic stability. In the case of entropy stabilization, a quasiperiodic ground state (structure at zero K) could not exist, QCs would be high-temperature (HT) phases. The crucial role of the VEC for the phase stability has been successfully used as a powerful tool for the discovery of new QCs [21]. Therefore it is hardly surprising, that intermetallic QCs and their ACs have been identified as Hume-Rothery-type phases (see, e.g., [4, 22, 23]. This supports the hypothesis of QCs being energy-stabilized. However, it does not exclude the possibility that in addition to electronic stabilization also an entropic contribution is necessary. Examples for HT Hume-Rothery phases are, for instance, the β- and δ-phases in the classical system Cu-Zn.

The hypothesis of QCs as entropy-stabilized HT phases (see, e.g. [24, 25]) cannot be ruled out based on the experimental evidence collected so far. Simple low-temperature (LT) experiments are not suitable for studying the LT stability of QCs due to the sluggish diffusion in intermetallics at LT. Attempts to force diffusion by LT ball milling induced a transition to a periodic phase in decagonal Al-Co-Ni [26]. In contrast, icosahedral Cd-Yb remained quasiperiodic under similar conditions [27].

QCs and their ACs differ slightly in their stoichiometries (see, e.g., [28] and Fig. 2). There is no lock-in transition QC$\leftrightarrow$AC known that would be comparable to those found for some incommensurately modulated structures (IMSs). A devil's staircase of ACs of increasing order does not exist either. It is also remarkable that only low-order ACs have been discovered so far. High-order ACs should have similar Gibbs free energies, G=H-TS, with similar H/TS ratios, as the related QC; the stabilizing mechanism for such a series of phases AC$\leftrightarrow$QC with slightly varying stoichiometry should be the same, consequently.

In the case of an IMS with displacive modulation, a continous change in the modulation wave

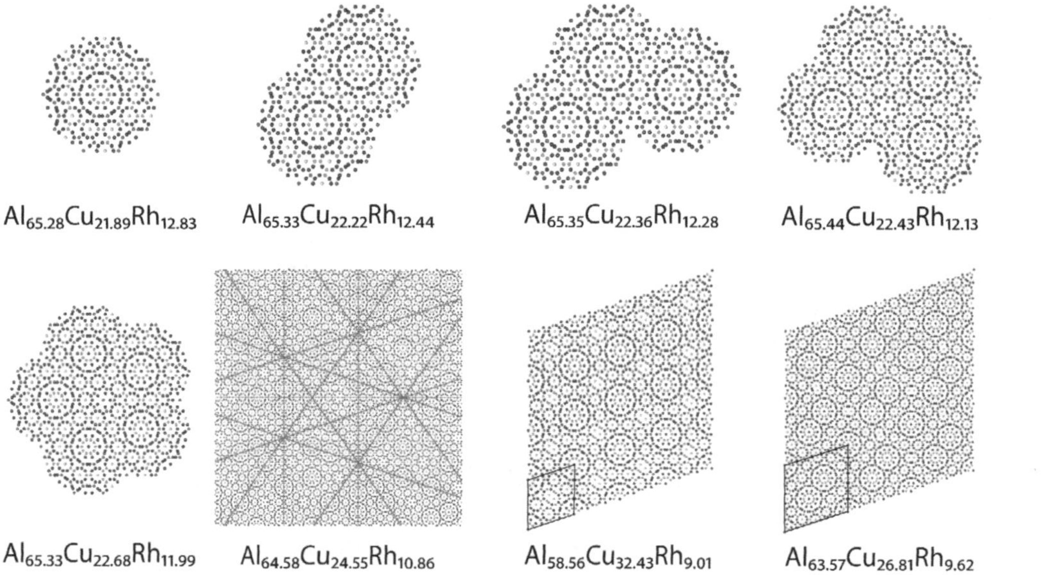

Figure 2. Arrangements of columnar clusters forming decagonal Al-Cu-Rh [9]. Their chemical composition varies with increasing size and the kind of cluster overlaps. The three structures at bottom right are patches of one quasiperiodic and two periodic arrangements of clusters, respectively. They differ significantly in their chemical composition. DQC growth could start from a pentagonal cluster (bottom left), growing along quasilattice lines as indicated schematically in the quasiperiodic structure.

vector would lead to a dense set of commensurate and incommensurate structures, all with the same stoichiometry. With temperature continously changing wave vectors have been experimentally observed, indeed. In contrast, in the case of a substitutional modulation, the stoichiometry would have to change with the wave vector, what would need diffusion, and what would have a larger influence on the Gibbs free energy. The same is true for a devil's staircase of ACs.

Since high-order ACs have never been experimentally observed so far, there seems to be a significant difference between their Gibbs free energies, indeed. This finding can be explained by the Hume-Rothery mechanism. Different stoichiometries lead to different valence electron concentrations and different structures. Those will be stable which allow the best possible fit of the respective Fermi surface to the Brillouin zone boundaries; the higher the symmetry is of a structure, the better the fit can be. For the series of approximants, the symmetry would always correspond to one of the 14 Bravais lattices. Consequently, the symmetry of teh QC would be highee. However, the Brillouin (Jones) zone of a high AC has already the pseudo-symmetry of the QC. Only if the size of the Fermi sphere (related to the electron concentration) fits best for the QC it will be stabler than any of its high ACs.

High-order ACs would have large unit cells. However, the unit-cell dimensions of periodic intermetallics seem to be quite limited. Only approximately 2% of all intermetallics have a complex structure ($>$ 100 atoms per primitive unit cell [5]) with larger lattice parameters. Thus, the structure stabilizing power of periodicity (Bloch waves, phonons,...) seems to work only for moderate unit cell dimensions (lattice parameters $<<$ 100 Å).

Open questions and ways to answer them
What are the main factors contributing to the stability of QCs? To which amount contribute phonons, phasons, electrons, and configurational disorder to the total entropy?

(i) Measure the entropies of a QC and its ACs in an intermetallic system by calorimetry (see, e.g., [29]. If the QC has the highest entropy it supports the hypothesis of QCs being HT phases.

(ii) Compare the variations in QC reflection intensities with high perp-space components of the diffraction vectors as a function of temperature (see, e.g. [30]). Correct the intensities for the par-space Debye-Waller factor. A decrease of the intensities with increasing temperature indicates an increase in phason fluctuations deteriorating the quasiperiodicity of the structures. An increase indicates the improvement of on average quasiperiodicity by decreasing the spontaneous local phason strains, precursor effects of a transition to the periodic state.

(iii) Perform first-principles calculations on a series of higher and higher ACs, 0/1, 1/1, 2/1, 3/2, $\cdots$, n/m. The results should explain why only low-order ACs have been found experimentally, and what changes as a function of the AC order Gibbs free energy, chemical bonding, electronic structure, etc.). The calculations should also allow to extrapolate the stability of the QC and the origin of quasiperiodic order - why are QCs quasiperiodic [31]?

4. QC growth

Not much is known about the growth of QCs on the atomic scale. This is not surprising because the growth mechanism of periodic complex intermetallic is also largely unexplored ('how does the thousandth atom find its place in a giant unit cell with thousands of atoms?'). In case of QCs, random tiling approaches [32] were quite popular in the first years of QC research, because, to say it with Roger Penrose's words "...the spontaneous growth of large regions of such quasicrystalline five-fold symmetric substances seemed to me virtually insurmountable" [33]. There are a couple of papers reporting approaches for modeling the growth of QCs based on tilings (see, e.g., [34] and references therein), and others based on clusters. From my point of view the cluster-based approaches are more physical (see, e.g., [35] and references therein). There are also a number of papers on experimental in situ studies such as [36, 37, 38] and references therein.

Open questions and ways to answer them

How do QCs and ACs grow on atomic/cluster scale? Do QCs already grow more or less strictly quasiperiodic or is this the result of optimization processes during annealing?

(i) Develop realistic 3D growth models, i.e., starting with nucleation and growth from a melt (see, e.g., [35]). Take into account that already in the melt, close to the melting temperature, chemical short-range ordering takes place. Depending on the stoichiometry either ACs should grow or QCs.

(ii) Study the heterogenous growth of QCs on surfaces of periodic crystals by AFM/STM; for instance, the growth of decagonal A-Co-Ni on the (110) surface of $cP2$-Al(Co,Ni) (B2- or β-phase) (for an example see [39].

(iii) Is it possible to grow a 2D QC on the surface of a periodic crystal?

References

[1] Shechtman D, Blech I, Gratias D and Cahn J W 1984 *Phys. Rev. Lett.* **53** 1951-3
[2] Steurer W and Deloudi S 2009 *Crystallography of Quasicrystals* (Berlin: Springer)
[3] Barkan K, Engel M and Lifshitz R 2014 *Phys. Rev. Lett.* **113** 098304
[4] Tsai A P 2004 *J. Non-Cryst. Sol.* **334 & 335** 317-22
[5] Steurer W and Dshemuchadse J 2016 *Intermetallics - Structures, Properties and Statistics* (Oxford University Press)
[6] Godrèche C and Lançon F 1992 *J. Phys. I France* **2** 207-20
[7] Ors T, Takakura H, Abe E and Steurer W 2014 *Acta Crystallogr. B* **70** 315-30
[8] Cervellino A, Haibach T and Steurer W 1998 *Phys. Rev.* B **57** 11223-31
[9] Kuczera P, Wolny J and Steurer W 2012 *Acta Crystallogr. B* **68** 578-89
[10] Steurer W 2006 *Philos. Mag.* **86** 1105-13
[11] Steurer W 2012 *Struct. Chem.* **23** 1115-20
[12] Hovmöller S, Hovmöller Zou L, Zou X and Grushko B 2012 *Phil. Trans. R. Soc. A* **370** 2949-59
[13] Deloudi S, Fleischer F and Steurer W 2011 *Acta Crystallogr. B* **67** 1-17
[14] Joseph D, Ritsch S and Beeli C 1997 *Phys. Rev.* B **55** 8175-83
[15] Hiraga K, Yasuhara A 2013 *Mater. Trans.* **54** 720-4
[16] Fan C Z, Weber T, Deloudi S and Steurer W 2011 *Philos. Mag.* **91** 2528-35
[17] Pavlovitch A and Kléman M 1987 *J. Phys. A: Math. Gen.* **20** 687-702
[18] Masáková Z, Patera J and Zich J 2003 *J. Phys. A: Math. Gen.* **36** 1869-94
[19] Masáková Z, Patera J and Zich J 2003 *J. Phys. A: Math. Gen.* **36** 1895-912
[20] Masáková Z, Patera J and Zich J 2005 *J. Phys. A: Math. Gen.* **38** 1947-60
[21] Tsai A P 2003 *Acc. Chem Res.* **36** 31-8
[22] Mizutani U, Inukai M, Sato H and Zijlstra E S 2012 *Chem. Soc. Rev.* **41** 6799-820
[23] Mizutani U, Inukai M, Sato H and Zijlstra E S, Lin Q 2014 *Philos. Mag.* **94** 2571-94
[24] De Boissieu M 2006 *Philos. Mag.* **86** 1115-22
[25] Kiselev A, Engel M and Trebin H R 2012 *Phys. Rev. Lett.* **109** 225502
[26] Hassdenteufel K H and Steurer W 2006 *Philos. Mag.* **86** 355-9
[27] Krauss G, Steurer W, Ross A R and LoGrasso T A 2006 *Philos. Mag.* **86** 505-11
[28] Ohhashi S, Hasegawa J, Takeuchi S and Tsai A P 2007 *Philos. Mag.* **86** 3089-94
[29] Holland Moritz D, Iu I R, Wilde G, Schroers J and Grushko B 1999 *J. Non-Cryst. Sol.* **252** 829-32
[30] Kuczera P, Wolny J and Steurer W 2014 *Acta Crystallogr. B* **70** 306-14
[31] Steurer W 2012 *Chem. Soc. Rev.* **41** 6719-29
[32] Joseph D and Elser V 1997 *Phys. Rev. Lett.* **79** 1066-9
[33] Penrose R in *Tilings and quasicrystals: a non-local growth problem?* 1989 Jaric M V (ed.) Academic Press Inc. London Ltd. **3** 53-79
[34] Hann C T, Socolar J E S and Steinhardt P J 2016 *Phys. Rev.* B **94** 014113
[35] Kuczera P and Steurer W 2015 *Phys. Rev. Lett.* **115** 085502
[36] Willmott P R, Schleputz C M, Herger R, Patterson B D, Hassdenteufel K and Steurer W 2005 *Phys. Rev.* B **71** 094203
[37] Gastaldi J, Reinhart G, Nguyen-Thi H, Mangelinck-Noel N, Billia B, Schenk T, Hartwig J, Grushko B, Klein H, Buffet A, Baruchel J, Jung H, Pino P and Przepiarzynskik B 2007 *Philos. Mag.* **87** 3079-3087
[38] Nagao K, Inuzuka T, Nishimoto K and Edagawa K 2015 *Phys. Rev. Lett.* **115** 075501
[39] Steurer W 2006 Z. *Kristallogr.* **221** 402-11

Cluster environments in a twelve-fold packing model of icosahedral quasicrystals

Hiroyuki Takakura[1] **and Radosław Strzałka**[2]

[1]Division of Applied Physics, Faculty of Engineering Hokkaido University, Sapporo, Hokkaido, 060-8628, Japan

[2]Faculty of Physics and Applied Computer Science, AGH University of Science and Technology, Mickiewicz Avenue 30, 30-059, Krakow, Poland

E-mail: takakura@eng.hokudai.ac.jp

Abstract. The shapes of the occupation domains that correspond to the respective local environments of the clusters within a 12-fold packing model of icosahedral quasicrystals (iQCs) have been derived for the first time. The frequencies of appearance of the local environments of the clusters have also been calculated based on the exact volumes. The present results validate the frequencies obtained by a previous study using a statistical approach. Application of the 12-fold packing model to the F-type Al-based iQCs is discussed briefly.

1. Introduction

The twelve-fold (12-fold) packing vertices is a subset of the 12-fold vertices that have all twelve possible edges radiating from them along 5-fold directions out of the vertices of a three-dimensional Ammann-Kramer-Neri (3D-AKN) tiling (or a 3D generalisation of Penrose tiling) [1, 2]. The 12-fold packing vertices are characterised by the shortest distance c between a pair of the vertices along 3-fold directions (c-link), and the next-shortest distance b along 2-fold directions (b-link). Their distances are related to a, the edge length of 3D-AKN tiling, as $b = a(4 + 8/\sqrt{5})^{1/2}$ for b-links and $c = b(\sqrt{3}/2)$ for c-links. Short a-links with length a that exist in the 12-fold vertices along 5-fold directions are eliminated. The geometries of the 12-fold packing vertices were studied in detail by Henley in the context of sphere packing problem as a theoretical structure model of icosahedral quasicrystals (iQCs) [2, 3]. A successful structural description for the Yb-Cd iQC has been given as a 6D structure model, where Tsai-type clusters are arranged at the 12-fold packing vertices generated by a particular shape of occupation domain (Fig. 1(b)) [4, 5]. The geometrical properties of iQCs model based on the 12-fold packing vertices was discussed in a literature [6].

The model of the Yb-Cd iQC is a representative model based on the 12-fold packing vertices of a 3D-AKN tiling, or 12-fold packing model for short [5, 6]. It consists of three building units: rhombic triacontahedron (RT), acute rhombohedron (AR), and obtuse rhombehedron (OR). By choosing an appropriate length for the edges for an underlying 3D-AKN tiling and by giving appropriate atomic decorations to the three building units, a complete atomic structure model for an iQC can be constructed. The binary Yb-Cd iQC model is the case that unique atomic decorations can be given to the building units [5, 6]. According to the previous studies, the 12-fold packing vertices have coordination numbers in the range from 7 to 12 [2, 6]. Frequencies

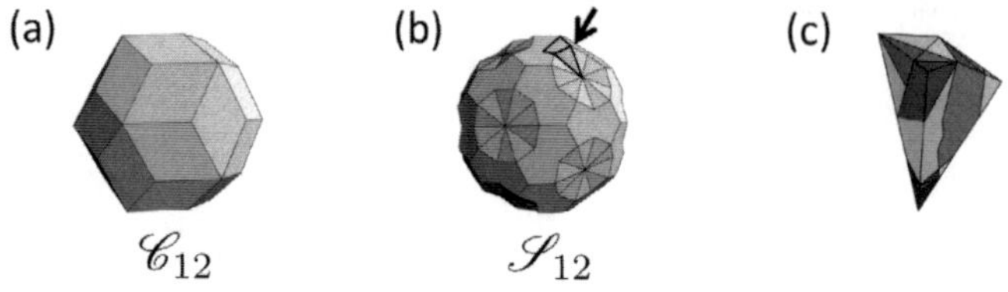

Figure 1. (a) The occupation domain $\mathscr{C}_{12}$ for the 12-fold vertices of a 3D-AKN tiling. (b) The archetype occupation domain $\mathscr{S}_{12}$ for the 12-fold packing vertices, and (c) its asymmetric part (arrowed part in (b)). The tessellation indicates the sectors that generate different vertex enviroments (see text and Fig. 2).

for 18 different vertex (or cluster) environments were evaluated numerically for the specific occupation domain (Fig. 1(b)) [6]. However, the corresponding shape of the occupation domain for each vertex environment is unknown. The aim of this work is to present the shapes of the occupation domains for each cluster environment for the specific occupation domain (Fig. 1(b)). This information will be useful when one consider an extended model of the Yb-Cd iQC including local chemistry for ternary iQCs or a similar model with different cluster types, such as Bergman or Mackay types or both.

2. Occupation domains

We recall the basic relationships between occupation domains defined in the internal space (or perpendicular space) within the section (or cut) method [1, 2, 7]. Consider an 3D-AKN tiling with edges of length a. The set of the vertices is generated by using an occupation domain, $\mathscr{C}$, with RT shape with edges of length a [1, 2]. The set of 12-fold vertices is a subset of those of the 3D-AKN tiling. The corresponding occupation domain, $\mathscr{C}_{12}$ (Fig. 1(a)), comes from the central part of $\mathscr{C}$, which also has RT shape yet with the edges of length a/τ^2 [2]. Here $\tau \equiv (1+\sqrt{5})/2$ is the golden mean. Since the 12-fold vertices include pairs of the vertices having the length a along 5-fold directions (a-links), one vertex form each pair has to be eliminated. The simplest realization of it is $\mathscr{S}_{12}$ (Fig. 1(b)). This occupation domain was utilised as an archetype occupation domain for the Yb-Cd iQC model [5].

3. Cluster environments and the shapes of occupation domains

The cluster local environment in a 12-fold packing model is distinguished by the numbers of b- and c-links and the configuration of them [2, 3, 6]. We use a similar notation that used in a previous study, such as (CN, Z_b, Z_c) and its corresponding representative 6D vector, where CN is a coordination number that is the sum of the number of b-links, Z_b, and that of c-links, Z_c [6]. For the sake of simplicity, the edge length for $\mathscr{C}_{12}$ is normalised as unity, namely $a/\tau^2 = 1$ in the following calculations.

In the present study, the exact shape of the sector corresponding to each local environment in the asymmetric part of $\mathscr{S}_{12}$ (Fig. 1(c)) was obtained by two steps. First, the shape of a sector was obtained by a program using CGAL: Computational Geometry Algorithms Library [8]. Second, the coordinates of the vertices for each sector were expressed in the unit of $(2+\tau)^{-1/2}$ and transformed them into error free formulae using $(\alpha + \beta\tau)/\gamma$, where α, β are any integers, and γ is a positive integer. The volume of each sector was calculated based on the vertices expressed with those error free formulae. The results are listed in table 1. The volume of the asymmetric part of $\mathscr{S}_{12}$ (Fig. 1(c)) is given as $(-37+23\tau)/6$. The frequency of appearance for each local environment is thus obtained simply as a relative ratio of the volumes. The values are given down to six decimal places. The present results validate the frequencies obtained by a previous study using a statistical approach [6].

The sectors that correspond to the respective local environments are shown in Fig. 2. Their shapes in the asymmetric part of $\mathscr{S}_{12}$ appear as tetrahedron, pentahedron, hexahedron, heptahedron, octahedron, hendecahedron, or dodecahedron. It should be noted that the sector of No.9 with (10,5,5) configuration and that of No.14 with (11,6,5) configuration consist of two disconnected parts, as indicated by arrows.

Table 1. Local environments in the twelve-fold packing vertices by $\mathscr{S}_{12}$ of the 3D-AKN tiling. The volume of the sector in the asymmetric part is presented in the unit of $(2+\tau)^{-1/2}$.

No.	CN	Z_b	Z_c	6D vector	Frequency (%)	Volume of the sector
1	7	0	7	(2 0 0 0 0 -1)	1.445942...	$(233-144\tau)/6$
2	8	1	7	(3 1 1 0 -2 0)	0.341340...	$(-987+610\tau)/6$
3	9	2	7	(4 1 1 0 0 -1)	0.421920...	$(1597-987\tau)/3$
4	9	5	4	(2 1 1 -2 0 -2)	1.787282...	$(-377+233\tau)/3$
5	10	3	7	(3 1 1 -1 -1 -2)	0.682681...	$(-987+610\tau)/3$
6	10	3	7	(5 1 0 1 0 0)	0.377376...	$(-7142+4414\tau)/15$
7	10	4	6	(5 1 0 0 0 0)	0.140640...	$(1597-987\tau)/9$
8	10	5	5	(3 1 0 -1 0 -2)	1.104601...	$(610-377\tau)/3$
9	10	5	5	(5 0 0 0 0 0)	1.642054...	$(-2804+1733\tau)/90$
10	10	6	4	(3 0 0 -1 0 -2)	10.46293...	$(322-199\tau)/3$
11	11	4	7	(4 1 0 -1 0 -1)	0.610608...	$(4414-2728\tau)/15$
12	11	5	6	(3 0 0 0 0 -3)	5.161148...	$(-631+390\tau)/18$
13	11	5	6	(4 0 0 -1 0 -1)	0.411660...	$(-1364+843\tau)/18$
14	11	6	5	(4 0 0 -1 -1 -1)	5.129700...	$(416-257\tau)/90$
15	11	6	5	(3 1 0 0 0 -3)	7.430412...	$(-1330+822\tau)/9$
16	12	5	7	(3 0 -1 -1 0 -2)	13.582381...	$(678-419\tau)/9$
17	12	6	6	(3 0 -1 -1 -1 -2)	24.398671...	$(-3857+2384\tau)/45$
18	12	7	5	(3 -1 -1 -1 -1 -2)	24.868639...	$(280-173\tau)/9$

4. Discussion

A cubic approximant crystal that corresponds to F-type Al-based iQCs in Al-Pd-Cr-Fe system has been found recently [9]. The structure can be interpreted as a 2 × 2 × 2 superstructure of Henley's cubic 3/2 packing, and that the parity of each vertex determines the kind of associated cluster, namely, pseudo-Mackay-type or mini-Bergman-type [3, 9]. Note also that the proper lattice constant estimated for F-type iQCs is $a \approx 0.56$ nm, which is τ^{-1} times smaller than that adopted by previous iQC models [10, 11]. It is known that a set of 12-fold packing vertices can be divided into even and odd parity vertices; the c-links have opposite parities, while the b-links have the same parities [2]. This discrimination results in F-type order with doubled unit cell in size in 6D space. By decorating each vertex with one of these two types of clusters according to the parity, a 6D model for F-type Al-based iQCs, such as i-Al-Pd-Mn and i-Al-Cu-Fe, can be constructed [12]. The pseudo-Mackay-type cluster is known to contain around 8 atoms, excluding a central atom, in the core of the cluster [13]. The knowledge of the local environment of the clusters is of help in understanding the number and the arrangement of those atoms geometrically, since the pseudo-Mackay-type cluster and mini-Bergman-type cluster heavily overlap with c-link, and the c-links are responsible for the formation of a pseudo-Mackay-type cluster [9, 12].

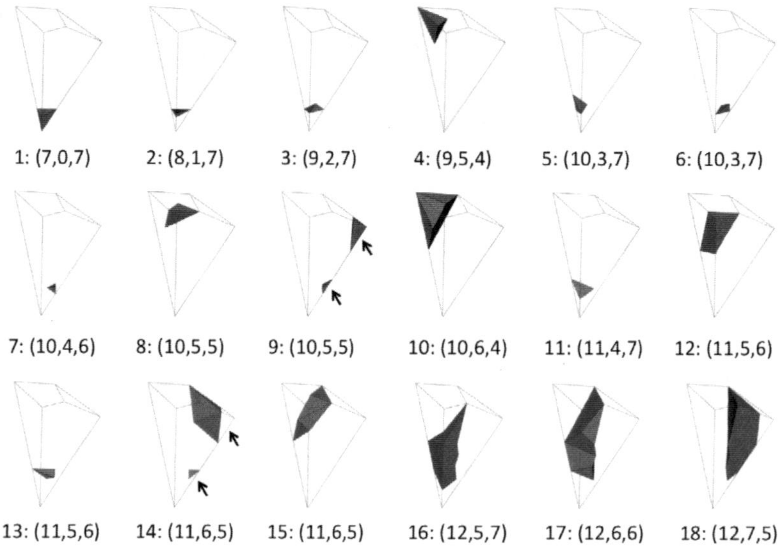

Figure 2. The sectors of the occupation domains that correspond to the respective local environments of the 12-fold packing vertices generated by $\mathscr{S}_{12}$.

5. Conclusion

In this paper we have derived the shape of the occupation domains that correspond to the particular local environments of clusters within a 12-fold packing model of iQCs. The frequencies of appearance of the particular local environments of clusters have also been calculated using the exact volumes. The present results validate the frequencies obtained by a previous study using a statistical approach [6]. Also the results will be useful when one consider an extended model of the Yb-Cd iQC including local chemistry for ternary iQCs or a similar model with different cluster types, such as Bergman-type or Mackay-type or both.

Acknowledgments

Authors kindly acknowledge the financial support by JSPS KAKENHI JP15K04659 (H.T.) and National Science Centre, Poland, under grant no. 2014/13/N/ST3/03776 (R.S.). The participation of R.S. in the project was possible thanks to agreement between AGH-UST Krakow and Hokkaido University.

References

[1] Kramer P, Neri R 1984 *Acta Cryst.* A **40** 580
[2] Henley C L 1986 *Phys. Rev.* B **34** 794
[3] Henley C L 1991 *Phys. Rev.* B **43** 993
[4] Tsai A P, Guo J Q, Abe E, Takakura H, and Sato T J 2000 *Nature* **408** 537
[5] Takakura H, Gómez C P, Yamamoto A, de Boissieu M and Tsai A P 2007 *Nat. Mater.* **6**, 58
[6] Takakura H 2008 *Phil. Mag.* **88** 1905
[7] Yamamoto A 1996 *Acta Cryst.* A **52** 509
[8] CGAL, The Computational Geometry Algorithms Library, http://www.cgal.org
[9] Fujita N 2013 *Acta Cryst.* A **69** 322
[10] Katz A, Gratias D 1993 *J. Non-Cryst. Sol.* **153&154** 187
[11] Yamamoto A, Takakura H and Tsai A P 2003 *Phys. Rev.* B **68** 094201
[12] Takakura H (*in prep.*)
[13] Boudard M, de Boissieu M, Janot C, Heger G, Beeli C, Nissen H -U, Vincent H, Ibberson R, Audier M and Dubois J M 1992 *J. Phys. Condens. Matter.* **4** 10149

IOP Conf. Series: Journal of Physics: Conf. Series **809** (2017) 012003 doi:10.1088/1742-6596/809/1/012003

Origin of 18-fold quasicrystal

S Bekku[1], P Ziherl[2,3] and T Dotera[1]

[1]Department of Physics, Kinki University, 3-4-1 Kowakae Higashi-Osaka 577-8502, Japan
[2]Faculty of Mathematics and Physics, University of Ljubljana, Jadranska 19, SI-1000 Ljubljana, Slovenia
[3]Jožef Stefan Institute, Jamova 39, SI-1000 Ljubljana, Slovenia

E-mail: dotera@phys.kindai.ac.jp

Abstract. We investigate the origin of the 18-fold symmetry in quasicrystals. We perform Monte Carlo simulations of 2D disks interacting with the hard-core/square-shoulder potential, and we analyze the bond-orientational order and the local vertex configurations. We find that the so-called special clusters of particles play a crucial role in the mixing of the three sets of bond directions 20° apart, producing the 18-fold diffraction pattern.

1. Introduction

Since soft quasicrystals were discovered in nature [1, 2, 3, 4], a new research front has emerged in the field of quasicrystals. In particular, Fischer *et al.* found that colloidal micelles in solution self-assembled into a 18-fold quasicrystalline phase [5], whose symmetry has not been found in synthetic metallic compounds [6, 7]. They used PI_{30}-PEO_{120} block copolymers based on the hydrophobic polyisoprene (PI) and the hydrophilic polyethylene oxide (PEO). These block copolymers self-assembled into spherical core-shell micelles where the well-defined core of PI blocks is surrounded by a thick shell of PEO blocks.Presently, the real-space structure of quasicrystals formed by these micelles is still not unequivocally determined.

To investigate soft quasicrystals theoretically, we have numerically explored a 2D system of disks with a purely repulsive hard-core/square-shoulder (HCSS) pair interaction [8, 9, 10, 11]

$$V(r) = \begin{cases} \infty, & r < \sigma \\ \varepsilon, & \sigma < r < \lambda\sigma \\ 0, & r > \lambda\sigma \end{cases} . \quad (1)$$

Here ε is the shoulder height and $\lambda\sigma$ and σ are the shoulder and the hard-core diameters, respectively. This two-lengthscale potential is known to stabilize several types of quasicrystals [12, 13, 14, 15, 16, 17], and its shape is described solely by the shoulder-to-core ratio λ.

At a shoulder-to-core ratio of $\lambda \approx 1.27$, we found a high-density phase of 18-fold symmetry [13, 18]. With a large fraction of particles forming close-packed equilateral triangles, the phase is structurally similar to the dodecagonal quasicrystalline phase except that the four-sided tiles that it contains are rhombi rather than squares. The rhombi, whose acute and obtuse angles are 80° and 100°, respectively, consist of two identical isosceles triangles such their bases are equal to $\lambda\sigma$ and their legs are σ. In this paper, we study the structure of the 18-fold phase in more detail, focusing on the mechanisms of its formation by examining the development of the bond-orientational order parameters at low temperatures.

2. Specific heat and bond-orientational order

We use Monte Carlo (MC) simulations of the NVT ensemble of $N = 4000$ particles in a square box with periodic boundary conditions. We choose the shoulder-to-core ratio of $\lambda = 1.27$ and a packing fraction of $\eta = \pi N\sigma^2/4A$, where N is the number of disks in a box of area A. In all results presented here except figure 10, $\eta = 0.77$. The value of $\lambda = 1.27$ was chosen such that the apex angle of the isosceles triangle mentioned above and given by $2\arcsin(\lambda/2)$ is close to 80°, thereby promoting the 80°-100° rhombi characteristic of the 18-fold quasicrystal. Inverse temperature is defined by $\beta = \epsilon/k_BT$. The system was equilibrated for 10^7 MC steps at all β between 0 and 5 except for $2.5 < \beta < 4.0$ where 5×10^7 MC steps were performed.

Figure 1 shows the specific heat per particle. The most striking feature of this plot is the sharp peak at $\beta = 3.1$, which indicates a transition. To understand the meaning of this peak, we evaluate the bond-orientational order parameters χ_m defined by

$$\chi_m = \left\langle \left| \frac{1}{N_B} \sum_{\mathbf{r}} \exp\left(im\theta_{\mathbf{r}}\right) \right|^2 \right\rangle ,$$

where $\theta_{\mathbf{r}}$ is the angle between a reference in-plane direction and the bond connecting particles separated by no more than $\lambda\sigma$ and located at $\mathbf{r}$; N_B is the number of nearest-neighbor bonds and angle brackets indicate ensemble average. Based on χ_6, χ_{12}, and χ_{18} shown in figure 2, one may distinguish between three distinct regimes. For $\beta < 3.1$, all three order parameters decrease with β and χ_6 is considerably larger than χ_{12} and χ_{18}, which indicates a 6-fold phase. The abrupt increase of χ_{18} at $\beta = 3.1$ consistent with the specific-heat peak signals the onset of the intermediate-temperature regime. The subsequent growth of χ_{18} and the continuing decrease of χ_6 and χ_{12} imply a coexistence of 18-fold and 6-fold phase for β between 3.1 and 3.9. The third, low-temperature regime begins at $\beta = 3.9$ where χ_{18} increases again and grows further as β is increased while χ_6 and χ_{12} are 0. This state corresponds to the pure 18-fold phase.

Figure 3 shows the diffraction patterns obtained by computing the Fourier transforms of particles' centers of mass at $\beta = 3.0, 3.2$, and 4.0, (panels a, b, and c, respectively). These

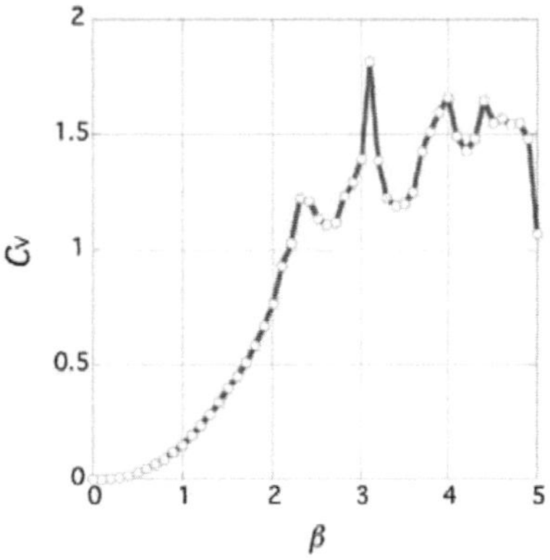

Figure 1. Specific heat C_V per particle *vs.* inverse temperature β; note the sharp peak at $\beta = 3.1$.

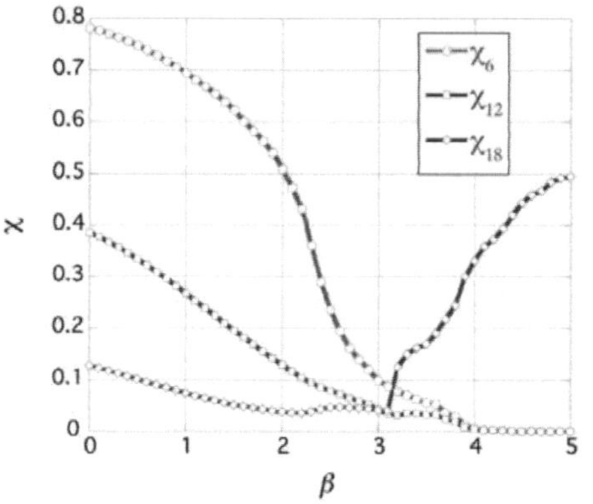

Figure 2. Bond orientational order parameters χ_6, χ_{12}, and χ_{18} *vs.* β. Beyond $\beta \cong 3.1$, χ_{18} increases whereas χ_6 and χ_{12} are both 0 for $\beta \geq 3.9$.

patterns additionally emphasize the difference between the three regimes, demonstrating that at $\beta = 4.0$ the particles form the 18-fold quasicrystal. Further development of the 18-fold diffraction pattern at lower temperatures still and the temperature dependence of the bond-order correlations are given in the supporting material of reference [13]. Here the nature of the 18-fold order is additionally elaborated by examining the local packing motifs.

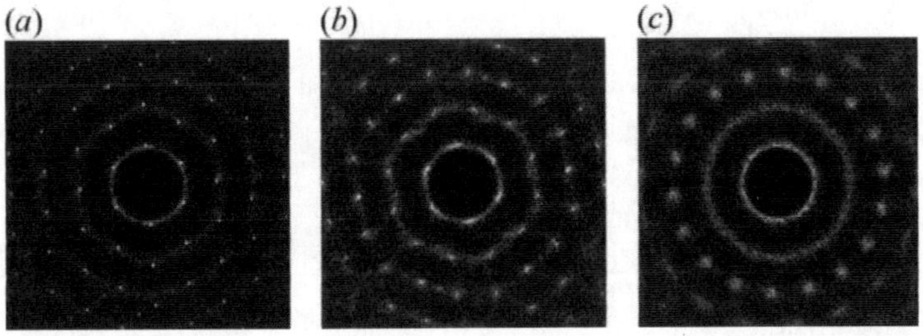

Figure 3. Fourier transforms of particles' centers of mass at $\beta = 3.0$ (a), 3.2 (b), and 4.0 (c) as representatives of the high-, intermediate-, and low-temperature regimes, respectively.

3. Frequencies of particle clusters

To this end, we classify the clusters of triangles and rhombi around a given vertex shown in figure 4. Each cluster is labeled by the corresponding vertex figure, a set of integers $(n_1.n_2.n_3.\cdots)$ that denotes the way that n_1-gon, n_2-gon, and n_3-gon, ... meet consecutively at each vertex [19]; and superscripts are employed to abbreviate the notation whenever possible.

We identify eight types of local clusters, which are more complex than those in the square-triangle dodecagonal tiling [20]. Three out of eight clusters (3^6, $3^2.4.3.4$, and $3^3.4^2$) are the same as in the square-triangle tiling whereas the 4^4_a and 4^4_b clusters are both generalizations of the same 4^4 cluster of squares. In the former, rhombi are arranged either such that their long diagonals all point in the same direction and in the latter they form a chevron pattern. These five clusters may be constructed from rhombi of arbitrary shape and are thus referred to as the generic clusters. On the other hand, the special clusters only exist in 80°-100° rhombi.

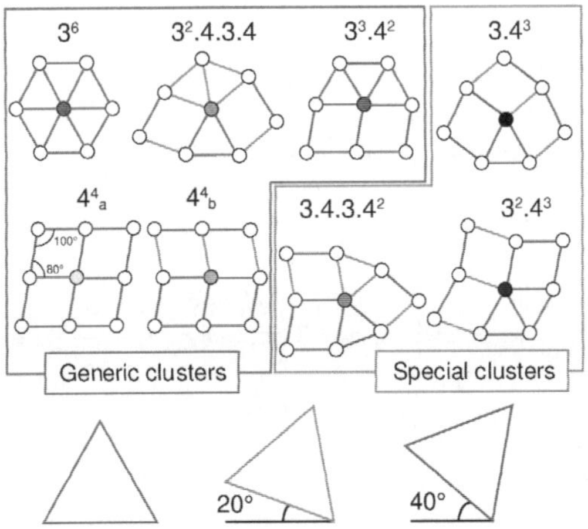

Figure 4. Classification of clusters in the 18-fold quasicrystal. The five generic clusters 3^6, $3^2.4.3.4$, $3^3.4^2$, 4^4_a, 4^4_b appear in any rhombus-triangle tiling, whereas the three special clusters 3.4^3, $3.4.3.4^2$, $3^2.4^3$ are characteristic of the 80°-100° rhombus-triangle tiling. The red, green, and blue triangles at the bottom represent the three sets of bond orientations. The special clusters and the 4^4_b cluster mix the three sets of orientations. The color code for vertices in the centers is used in figures 5, 7, and 8.

These clusters are needed so as to integrate the chevron cluster 4_b^4 into the matrix; the 3.4^3 accommodates its recessed side whereas the $3.4.3.4^2$ and $3^2.4^3$ accommodate its pointed side.

In figure 4, the edges appearing in the eight clusters are colored according to orientation. The 18-fold quasicrystal is marked by 18 bond directions which can be divided into three sets of 6 directions 60° apart and rotated with respect to each other by 20°. The three sets are conveniently represented by the red, green, and blue equilateral triangles drawn at the bottom of figure 4. The edge coloring shows that all generic clusters except 4_b^4 contain only orientations from two of these sets whereas the 4_b^4 cluster and all special clusters mix all three sets of bond orientations. This shows why they are key to understand the formation of 18-fold quasicrystals.

The frequencies of these eight cluster types is presented in figure 5, the special clusters all combined in a single class for clarity; also plotted is the frequency of all clusters beyond this classification. In a triangular tile, the overlap energy per particle of triangles is larger than in a rhombus. This explains why the frequency of the triangle-only 3^6 clusters decreases on cooling, whereas the frequencies of the 4_a^4 and 4_b^4 clusters is finite rather than 0 in the intermediate- and low-temperature regime at $\beta > 3.1$. On cooling towards $\beta = 3.1$, the frequencies of the 3^6 steadily decreases whereas those of the $3^2.4.3.4$ and the $3^3.4^2$ clusters increase. This is consistent with a decrease of the size of hexagonal domains. At the onset of the intermediate-temperature regime at $\beta = 3.1$, the frequencies of the $3^2.4.3.4$ and the $3^3.4^2$ clusters jump abruptly at the expense of the 3^6 clusters, which drops dramatically.We note that many local clusters do not belong to one of the 8 types in figure 4, and these are referred to as "other clusters" in figure 5.

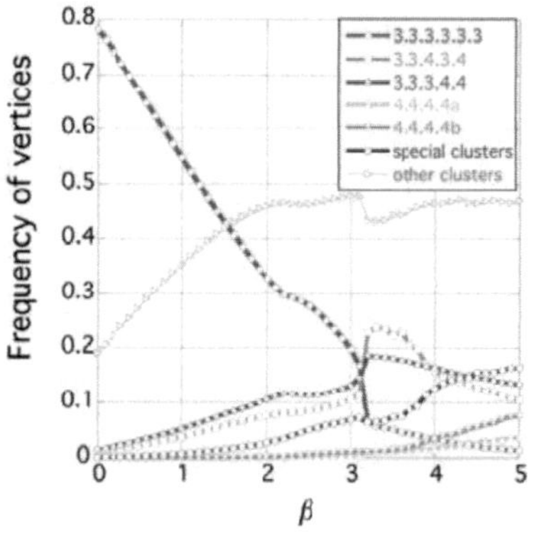

Figure 5. Frequencies of cluster types *vs.* β; special and non-classified ("other") clusters are plotted as single classes for clarity. At $\beta = 3.1$, abrupt changes of frequencies of several cluster classes are observed.

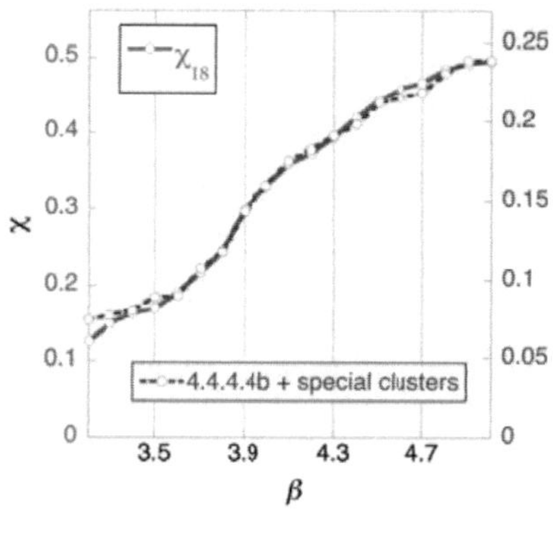

Figure 6. Superimposed 18-fold bond-orientational order parameter χ_{18} and the combined frequency of special and 4_b^4 clusters adapted from figures 2 and 5, respectively.

The gradual decrease of the frequencies of the $3^2.4.3.4$ and $3^3.4^2$ clusters and the concomitant increase of the frequency of special clusters on cooling beyond $\beta = 3.2$ are due to the different overlap energies of these clusters. All special clusters contain 3 rhombi whereas the $3^2.4.3.4$ and $3^3.4^2$ clusters contain only 2, and thus the former have a lower overlap energy than the latter. Also characteristic for the 2D 18-fold phase is a large frequency of clusters beyond the classification in figure 4, typically just a little short of 50 % as shown in figure 5.

These results all suggest that the onset of the 18-fold order parameter is closely correlated with the occurrence of the three special clusters, which are in turn needed for the integration

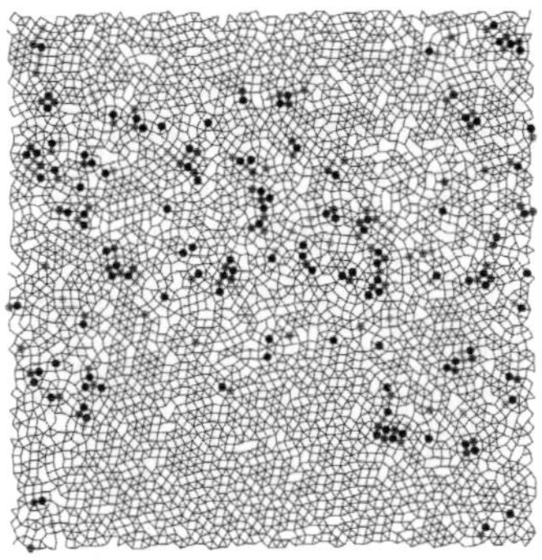

Figure 7. Coexistence of hexagonal and 18-fold phase at $\beta = 3.2$: The 4_b^4 and special clusters, which are color-coded like in figure 4, are not distributed uniformly.

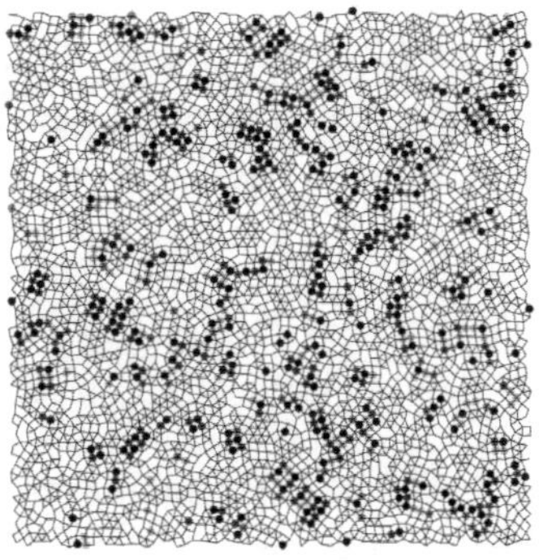

Figure 8. Spatial distribution of 4_b^4 and special clusters in the pure 18-fold phase at $\beta = 4.0$ is essentially uniform.

of the 4_b^4 clusters into the tiling. This correlation is best appreciated by plotting χ_{18} *vs.* β together with the combined frequencies of these four cluster types (figure 6), which shows that the two quantities are proportional to each other. The reason for this correlation is clear in representative snapshots of the system at $\beta = 3.2$ and $\beta = 4.0$ in figures 7 and 8, respectively. Here the clusters that mix the three sets of bond orientations are indicated by colored circles with yellow, sky blue, dark blue, and black corresponding to 4_b^4, $3.4.3.4^2$, $3^2.4^3$, and 3.4^3 clusters, respectively. At $\beta = 3.2$ these clusters are present in the center band-region but not elsewhere, which is characteristic of phase coexistenceexpected in a constant-volume ensemble. At $\beta = 4.0$ they are distributed uniformly and this makes a complete 18-fold diffraction pattern. In both cases, many 4_b^4 clusters form short chains typically terminated by special clusters. We conclude that the frequency of the special clusters is proportional to the area occupied of the 18-fold region.

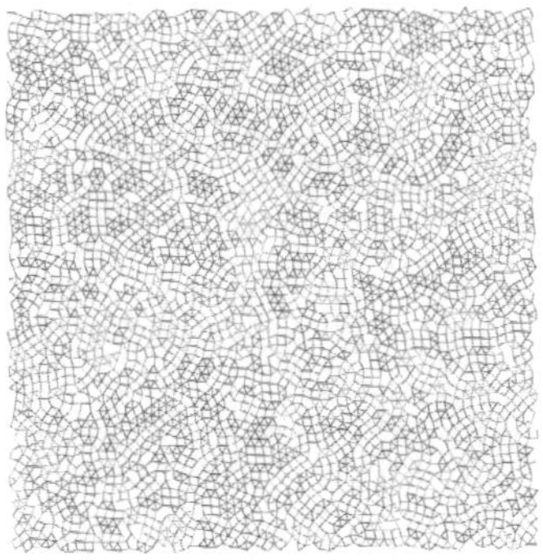

Figure 9. Snapshot from figure 8 redrawn by coloring the bonds according to the scheme defined at the bottom of figure 4. All three sets of directions are present.

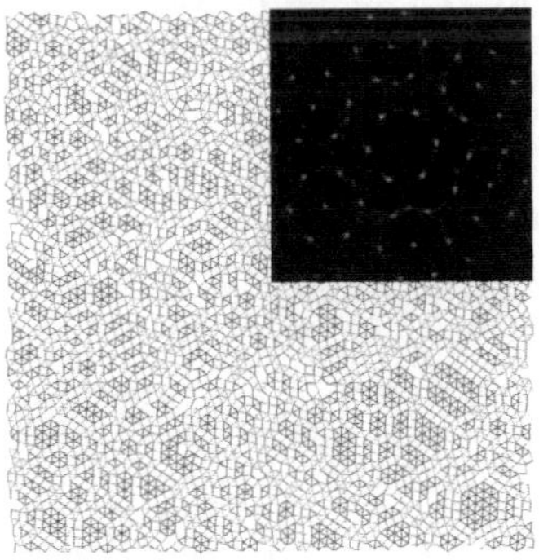

Figure 10. Colored-bond representation of the hexagonal phase at $\beta = 3.4$ and a packing fraction $\eta = 0.79$. Fourier transform emphasizes the 6-fold symmetry.

In figure 9, figure 8 is replotted by coloring the edges according to their orientation using the color scheme introduced by the three bond-direction triangles in figure 4. At $\beta = 4.0$, the three sets of directions are well mixed such that any small patch of the system has an 18-fold symmetry. On close inspection, we see that the top and the bottom parts of figure 7 devoid of special clusters primarily consist of hexagonal domains separated by domain boundaries based on $3^3.4^2$ clusters. As a result, the hexagonal phase contains only two sets of bond orientations. This is more clearly seen in the system at a slightly larger packing fraction of $\eta = 0.79$ and a somewhat higher temperature of $\beta = 3.4$ where the stable state is purely hexagonal (figure 10).

4. Discussion

The above results elucidate the role of special clusters in the formation of the 18-fold quasicrystal, showing that they ensure that the three sets of bond orientations are equiprobable.Interestingly, two of our special clusters are also present in the dissection of a regular decaoctagon into 18 pentagons with internal angles of $60°, 160°, 80°, 100°$ and $140°$ provided that the pentagon is viewed as combination of an equilateral triangle and a $80° - 100°$ rhombus (figure 11a). This dissection can be admired in the intricate pattern on the facade of the Ravensbourne College in London, designed by Foreign Office Architects. The pattern appears to be derived from the 6-fold $18^2.3^*_{2\pi/9}$ decaoctagon-three-pointed star tiling [19] by shrinking the decaoctagons and dissecting them into the $60° - 160° - 80° - 100° - 140°$ pentagons; the gaps between them are dissected into pentagons, triangles, and concave pentagons (figure 11b). The facade motif may be viewed as a hexagonal approximant of the 18-fold quasicrystal (plane group P6).

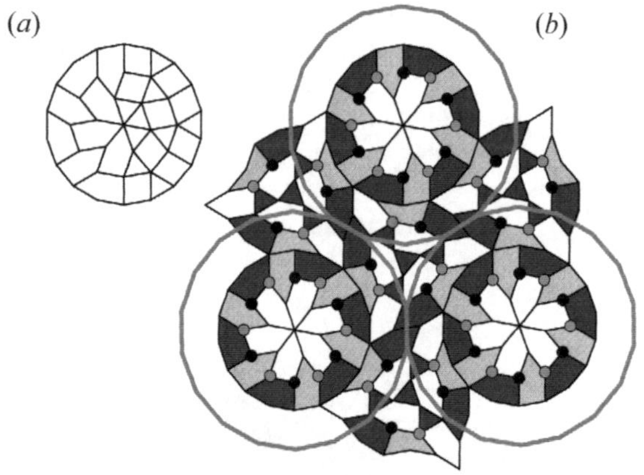

Figure 11. Dissection of the regular decaoctagon into irregular pentagons (a, left portion) which can be further divided into equilateral triangles and $80° - 100°$ rhombi (a, right portion). Motif from the Ravensbourne College (b). The red contours show a part of the underlying 6-fold $18^2.3^*_{2\pi/9}$ decaoctagon-three-pointed star tiling and its dissection. Also highlighted are the two special clusters [$3.4.3.4^2$ (sky blue) and 3.4^3 (black)] present in the motif.

Acknowledgments

This work was supported by Japan Society for the Promotion of Science [Grant-in-Aid for Scientific Research (B) No. 16H04037], by Slovenian Research Agency (Grant No. P1-0055), and by the Marie-Skłodowska-Curie European Training Network COLLDENSE (H2020-MSCA-ITN-2014 Grant No. 642774).

References

[1] Zeng X, Ungar G, Liu Y, Percec V, Dulcey A E and Hobbs J K 2004 *Nature* **428** 157-160
[2] Hayashida K, Dotera T, Takano A and Matsushita Y 2007 *Phys. Rev. Lett.* **98** 195502
[3] Lee S, Bluemle M J and Bates F S 2010 *Science* **330** 351-353
[4] Dotera T 2011 *Isr. J. Chem.* **51** 1197-1205
[5] Fischer S, Exner A, Zielske K, Perlich J, Deloudi S, Steurer W, Lindner P and Förster S 2011 *Proc. Natl. Acad. Sci. USA* **108** 1810-1814

[6] Janssen T, Chapuis G and de Boissieu M 2007 *Aperiodic crystals: From modulated phases to quasicrystals* (Oxford: Oxford University Press)
[7] Steurer W & Deloudi S 2009 *Crystallography of quasicrystals: Concepts, methods and structures* (Berlin: Springer)
[8] Young D A and Alder B J 1977 *Phys. Rev. Lett.* **38** 1213-1216
[9] Jagla E A 1999 *J. Chem. Phys.* **110** 451-456
[10] Malescio G and Pellicane G 2003 *Nat. Mater.* **2** 97-100
[11] Glaser M A, Grason G M, Kamien R D, Košmrlj A, Santangelo C D and Ziherl P 2007 *Europhys. Lett.* **78** 46004
[12] Barkan K, Diamant H and Lifshitz R 2011 *Phys. Rev. B* **83** 172201
[13] Dotera T, Oshiro T and Ziherl P 2014 *Nature* **506** 208-211
[14] Archer A J, Rucklidge A M and Knobloch E 2015 *Phys. Rev.* E **92** 012324
[15] Pattabhiraman H, Gantapara A P and Dijkstra M 2015 *J. Chem. Phys.* **143** 164905
[16] Ryltsev R, Klumov B and Chtchelkatchev N 2015 *Soft Matter* **11** 6991-6998
[17] Schoberth H G, Emmerich H, Holzinger M, Dulle M, Förster S and Gruhn T 2016 *Soft Matter* **12** 7644-7654
[18] Engel M and Glotzer S C 2014 *Nat. Phys.* **10** 185-186; Fig. 1 in this reference contains an artistic interpretation of the 18-fold quasicrystal by A. Šiber.
[19] Grünbaum B and Shephard G C 2016 *Tilings and patterns*, 2nd Edition (New York: Dover)
[20] Oxborrow M and Henley C L 1993 *Phys. Rev. B* **48**, 6966-6998

Crystal structure of χ-AlNiRu containing pseudo-Mackay Clusters

S Suzuki[1], R Simura[1], A Yoshiasa[2], and K Sugiyama[1]

[1] Institute for Materials Research (IMR), Tohoku University, Sendai 980-8577, Japan
[2] Faculty of Science, Kumamoto University, Kumamoto 860-8555, Japan

The corresponding author's e-mail address: kazumasa@imr.tohoku.ac.jp

Abstract. χ-AlNiRu (H-AlNiRu), as determined by single crystal X-ray diffraction, crystallizes in the trigonal space group, *P*31*c* (No.159), with unit cell parameters of $a = 1.2095(1)$ nm, $c = 2.6879(1)$ nm, and $V = 3.4056(3)$ nm^3, and atoms/cell = 208.6. The refinement converged with $R(F) = 0.087$ for 3915 observed reflections measured using Mo $K\alpha$ radiation ($\lambda = 0.071073$ nm). The structure of χ-AlNiRu is similar to that of χ-AlPdRe, wherein pseudo-Mackay type clusters are formed by the disordered distribution of Al in the heavy metal framework.

1. Introduction

Pseudo-Mackay clusters (pMC) are a common structural unit of quasicrystalline approximants, and a variety of linkages between pMCs have been described in previous reports [1–5]. For the ternary Al-Ni-Ru system, two types of decagonal phases (D-phases) have been reported, and high-resolution transmission electron microscopy and large-angle annular detector dark-field scanning transmission electron microscopy have been used for obtaining structural images of D-phases [6, 7]. Furthermore, an H-phase (hereafter, denoted χ-AlNiRu) with a chemical composition similar to that of the D-phase has been reported [8]. Herein, we report pMCs and their linkages within the structure of χ-AlNiRu. The Al-Ni-Ru system indicates the decagonal approximants of monoclinic and orthorhombic and cubic phases [9-11]. In comparison with the D-phases, an icosahedral AlNiRu phase has also been reported in the Ni-rich region [12].

2. Experimental

An alloy ingot with a composition of $Al_{77}Ni_{16}Ru_7$ was prepared by conventional arc melting in a purified Ar atmosphere. Fragments of the as-prepared ingot were charged into a carbon crucible, and the crucible was placed in an evacuated silica tube. The fragments were first heated to 1100 °C, subsequently cooled to 800 °C, and annealed at this temperature for 5 h. The annealed sample was mainly composed of χ-AlNiRu phase. Electron probe microanalysis (EPMA: JEOL JXA-8621MX) showed that the average chemical composition of χ-AlNiRu was $Al_{75.9}Ni_{16.2}Ru_{8.2}$. A small amount of a crystalline $Al_3Ni(Al_{74.1}Ni_{23.6}Ru_{2.1})$ phase was also found in the annealed sample. A single-crystal specimen of χ-AlNiRu was cut from the annealed sample and used for single-crystal X-ray diffraction analysis. Intensity data for 12997 reflections ($-8 \leq h \leq 15$, $-15 \leq k \leq 10$, $-34 \leq l \leq 34$) were collected using a Rigaku R-AXIS RAPID system equipped with an imaging plate. Data reduction, assuming trigonal *P*31*c* symmetry yields 5187 reflections ($R_{int}(I) = 0.0394$), among which 3915 reflections meet the condition of $F_o > 4.0\sigma(F_o)$. Lattice constants of $a = 1.2095(1)$ nm and $c = 2.6879(1)$ nm were determined by least-squares method, as implemented in the RAPID-AUTO system. Absorption correction was carried out using an integration method, assuming an arbitrary shape for the specimen. An initial structural model for χ-AlNiRu was obtained by direct methods using SIR97 [13]. Subsequently, the structure refinement was carried out by least-squares method (SHELXL-97) [14].

Table 1. Fractional coordinates of trigonal χ-AlNiRu.

Atom	Site	Occupation	x	y	z	$U_{eq.}$ ($U_{iso.}$)
Ru1	6*c*	1	0.5383(2)	0.4614(2)	0.1615(2)	0.0123(5)
Ni/Ru2	6*c*	0.86/0.14(3)	0.1196(3)	0.2397(3)	0.3768(2)	0.0118(11)
Ni/Ru3	6*c*	0.88/0.12(3)	0.4088(3)	0.2045(3)	0.0283(2)	0.0073(11)
Ni/Ru4	2*a*	0.94/0.06(4)	0	0	0	0.0057(12)
Ni5	6*c*	0.51(5)*	0.2373(9)	0.103(2)	0.2496(3)	0.0093(6)*
Ni5′	6*c*	0.49(5)*	0.2366(10)	0.134(2)	0.2495(3)	0.0093(6)*
Ru/Ni6	2*b*	0.86/0.14(5)	1/3	2/3	0.1033(3)	0.0165(13)
Ru/Ni7	6*c*	0.58/0.42(4)	0.4097(3)	0.2051(3)	0.4683(2)	0.0214(11)
Ni8	6*c*	1	0.2272(4)	0.4544(1)	0.2481(3)	0.0088(4)
Ni/Ru9	2*b*	0.51/0.49(5)	1/3	2/3	0.3937(3)	0.0102(15)
Ru/Ni10	6*c*	0.74/0.26(4)	0.5389(2)	0.4607(2)	0.3355(2)	0.0084(8)
Ni11	6*c*	1	0.1201(3)	0.2405(3)	0.1207(2)	0.0131(7)
Al1	2*a*	1	0	0	0.2496(6)	0.0168(17)
Al2	6*c*	1	0.1622(7)	0.0810(7)	0.1638(3)	0.0095(13)
Al3	6*c*	1	0.4274(8)	0.2127(7)	0.3048(4)	0.0138(17)
Al4	6*c*	1	0.0027(8)	0.3892(7)	-0.0005(3)	0.0172(14)
Al5	6*c*	1	0.3864(7)	0.3890(7)	-0.0005(3)	0.0166(14)
Al6	6*c*	1	0.2156(7)	0.4311(7)	0.0616(4)	0.0135(13)
Al7	6*c*	1	0.3465(8)	0.2891(8)	0.1035(3)	0.0163(16)
Al8	6*c*	1	0.0891(7)	0.5442(7)	0.0879(4)	0.0234(15)
Al9	6*c*	1	0.3465(7)	0.0583(8)	0.1034(3)	0.0166(17)
Al10	6*c*	1	0.3466(7)	0.2936(7)	0.3949(3)	0.0108(14)
Al11	6*c*	1	0.3462(7)	0.0539(7)	0.3950(3)	0.0120(15)
Al12	6*c*	1	0.115(1)	0.234(2)	0.2211(6)	0.075(4)
Al13	6*c*	1	0.1636(8)	0.0818(9)	0.3366(4)	0.0204(17)
Al14	2*b*	1	1/3	2/3	0.0140(12)	0.092(9)
Al15	6*c*	0.50(1)*	0.1588(14)	0.4543(14)	0.1591(6)	0.026(2)*
Al15′	6*c*	0.50(1)*	0.2946(14)	0.4529(14)	0.1593(6)	0.026(2)*
Al16	6*c*	1	0.5894(7)	0.4120(7)	0.0689(3)	0.0083(13)
Al17	2*b*	1	2/3	1/3	0.1523(5)	0.009(2)
Al18	2*b*	1	2/3	1/3	0.3435(5)	0.016(3)
Al19	6*c*	1	0.5900(8)	0.4114(8)	0.4265(3)	0.0171(16)
Al20	6*c*	1	0.5866(9)	0.4129(9)	0.2482(4)	0.0105(8)
Al21	6*c*	1	0.4468(8)	-0.1063(8)	0.3865(4)	0.0421(19)
Al22	6*c*	1	0.4228(10)	0.2106(9)	0.1892(4)	0.027(2)
Al23	2*a*	0.77(2)*	0	0	0.0879(5)	0.010(2)*
Al23′	6*c*	0.23(3)*	0.156(3)	0.063(3)	0.0660(10)	0.010(2)*
Al24	6*c*	1	0.0403(9)	0.3375(9)	0.3020(4)	0.0385(19)
Al25	6*c*	1	0.2960(9)	0.3370(9)	0.3020(4)	0.0383(19)
Al26	6*c*	1	0.4590(9)	0.5406(9)	0.2560(4)	0.0365(19)
Al27	6*c*	0.60(1)*	0.1587(13)	0.0731(13)	0.4337(5)	0.0222(19)*
Al28	6*c*	0.60(1)*	0.2078(14)	0.4143(13)	0.4354(6)	0.0222(19)*
Al29	2*b*	1	1/3	2/3	0.2993(14)	0.111(12)
Al30	2*b*	0.47(2)*	1/3	2/3	0.4877(9)	0.015(2)*
Al30′	6*c*	0.53(2)*	0.5859(11)	0.4135(11)	-0.0247(5)	0.015(2)*
Al31	2*b*	1	1/3	2/3	0.2012(7)	0.044(4)
Al32	6*c*	0.40(1)*	0.1113(16)	0.2230(17)	0.4708(7)	0.0222(9)

*See text

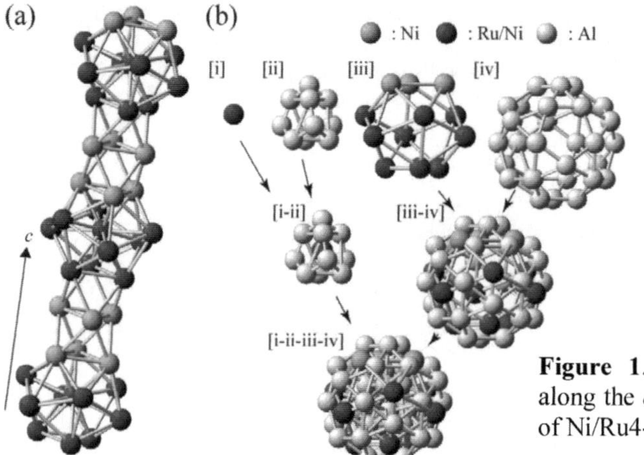

Figure 1. (a) The linkage of Ni/Ru4-pMC along the *c*-axis and (b) detailed shell structure of Ni/Ru4-pMC.

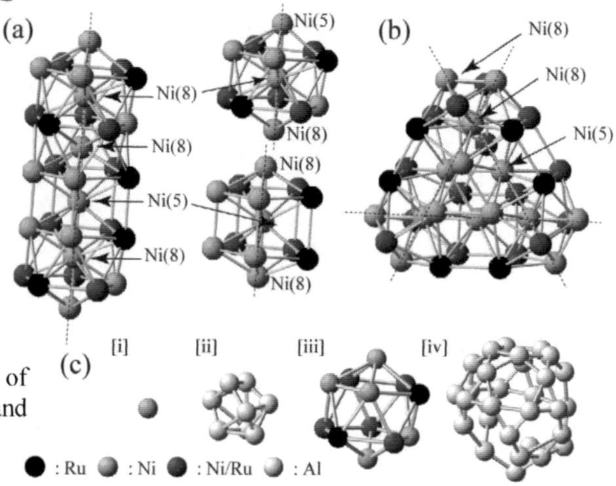

Figure 2 (a) The columnar structure of heavy metal polygons, (b) i3 cluster, and (c) imperfect Ni8-pMC.

The obtained structural model indicates a similar atomic arrangement to that of χ-AlPdRe [15], yielding 47 independent atomic sites per unit cell. Ten heavy metal sites and thirty-one Al are fully occupied by the respective elements, whereas the other sites are either partially occupied or disordered. After several least-squares iterations, the Fourier maps indicated the location of the possible paired sites at Ni5/Ni5′, Al15/Al15′, Al23/Al23′, Al27/Al32, Al28/A32, and Al30/Al30′. Because the short calculated distances exclude simultaneous occupation of these paired sites, we introduced a restriction that the sum of the occupancies for each pair is 1.0. During the structural refinement, the isotropic temperature factors for the paired Al sites were constrained to be equal. After modeling the disorder and refining the fully occupied heavy metal sites anisotropically, the refinement finally converged to an *R*(*F*) of 0.086 for all observed reflections. The maximum and minimum residual electron densities are 4.66 and -2.24 e$Å^{-3}$, respectively. The obtained chemical composition $Al_{74.1}Ni_{17.1}Ru_{8.8}$ is consistent with $Al_{75.9}Ni_{16.2}Ru_{8.2}$, as determined by EPMA. The final atomic coordinates and isotropic or equivalent temperature factors are listed in Table 1. The calculated atomic distances are reasonable except for the previously mentioned paired sites.

3. Results

The distribution of heavy metal sites of χ-AlNiRu can make a unique substructure with space group symmetry $P6_3/mmc$. Nevertheless, the remaining Al sites reside in the interstitial space of the substructure, resulting in the symmetry reduction toward subgroup symmetry $P31c$. This feature in the mode of elemental distribution leads to a local structure without direct bonding between heavy metals in the first nearest neighbor region. Notably, the substructure composed of heavy metal sites is isotypic with the structures of χ-AlPdRe and Ir_9Al_{28} [16] and significant structural differences are only present in the distribution of Al sites. Figure 1(a) shows the connected heavy metal icosahedra around Ni/Ru4 along the *c*-axis via the Ni5 or Ni5′ triangles, and a pMC with a heavy metal center is realized around Ni/Ru4. As shown in Fig. 1(b), the first irregular coordination shell is composed of disordered Al sites Al23, Al23′, Al27, and Al32. The second shell (Fig. 1(b-iii-iv))) may be grouped into two sub-shells. An icosahedron is produced by Ni/Ru2, Ni/Ru3, Ru/Ni7, and Ni11 and each vertex of the icosahedron is coordinated by five outer Al atoms, forming Al-icosidodecahedrons. It may be added that similar pMC with a disordered inner shell are also found in the approximants of *R*-AlPdCo [1] and χ-AlPdRe. Figure 2(a) shows the substructure units of Ni8-centered icosahedrons and Ni5(Ni5′)-centered bi-capped pentagonal prisms composed of heavy metals. The Ni8-centered icosahedra interpenetrate, producing an I-type connection (Fig. 2(b)) [17]. Such I-type connections are further connected via a common vertex at Ni5 (Ni5′) forming a pentagonal columnar structure along the *a*-axis. Figure 2(c) shows the Al distribution of the inner and outer area of the Ni(8)-centered icosahedrons. The first irregular coordination shell is composed of Al12, Al15′, Al15′, Al24, Al25, Al26, Al29, and Al31, similar to the case of the pMC around Ni/Ru4. The second shell is ranked into sub-shells of a heavy metal icosahedron and an imperfect Al-icosidodecahedron. The heavy metal icosahedron is composed of Ru1, Ni/Ru2, Ni5 (Ni5′), Ru/Ni6, Ni8, Ni/Ru9, Ru/Ni10, and Ni11 (Fig. 2(c-iii)). Interestingly, the pseudo five-fold axes along the neighboring Ni8 is broken in the outer Al-icosidodecahedron. Nevertheless, the distribution of Al in other directions indicates some similarities with Mackay cluster geometry.

4. Conclusion

The crystal structure of χ- AlNiRu was determined using single-crystal X-ray diffraction. The Heavy metal distribution corresponds to that in χ-AlPdRe and Ir_9Al_{28}, wherein the Al sites at the interstitial spaces of heavy metals are disordered. The atomic arrangement of the heavy metals is well featured by the unit structure composed of Ni/Ru4 and Ni8-centered pMCs. In particular, the Ni8-pMC shows the linkage of pMCs toward the five-hold direction running parallel to the *a*-axis and an aggregation of three pMCs in the I3 cluster. The knowledge of these structural units and their linkage would provide an image leading the structure of D- and I-phases of the AlNiRu system.

References

[1] Simura R, Suzuki S, Yubuta K, and Sugiyama K 2013 *Mater. Trans. JIM* 54 1385.
[2] Sugiyama K, Yubuta K, Yokoyama Y, Suzuki S, and Simura R 2013 *Acta Physica Polonica* A126 588.
[3] Mahne S and Steurer W 1996 *Z. Kristallogr.* 211 17.
[4] Sugiyama K, Kato T, Saito K, and Hiraga K 1998 *Philos. Mag. Let.* 77 165.
[5] Sugiyama K, Hiraga K, and Saito K 2000 *Mater. Sci. Eng.* A294 345.
[6] Sun W and Hiraga K 2001 *Philos. Mag. Lett.* 81 187.
[7] Sun W, Ohsuna T, and Hiraga K 2001 *Philos. Mag. Lett.* 81 425.
[8] Mi S, Grushko B C.Dong, and Urban K 2004 *J. Non-Cryst. Solid* 335 214.
[9] Murao R, Genba M, Sugiyama K, and Sun W 2011 *Mater. Trans. JIMs* 52 1344.
[10] Suzuki S, Murao R. Simura R, and Sugiyama K 2012 *Z. Kristalogr. NCS* 227 1.
[11] Hohls J, Cornish L A, Ellis P, and Witcomn M J 2000 *J. Alloy Compd.* 308 205.
[12] Wei S and Hiraga K 2002 *J. Alloy. Compd.* 347 110.
[13] Altomare A, Cascarano G, Giacovazzo C, Guagliarde A, Burla M C, Polidori G, and Camalli M 1999 *J. Appl. Crystallogr.* 32 115.
[14] Sheldrick G M 2008 *Acta Crystallogr.* A64 112.
[15] Suzuki S, Simura R, and Sugiyama K 2013 *Philos. Mag.* 91 2610.
[16] Katrych S, Gramlich V, and Steurer W 2006 *J. Alloy. Compd.* 407 132.
[17] Kreiner G and Franzen H F 1995 *J. Alloy. Compd.* 221 15.

Medium Range Ordering realized in $Zr_{80}Pt_{20}$Amorphous Alloy

K Sugiyama, T Kawamata and T Muto

Institute for Materials Research (IMR), Tohoku University, Sendai 980-8577, Japan

The corresponding author's e-mail address: kazumasa@imr.tohoku.ac.jp

Abstract. The short- and medium-range ordering structure in $Zr_{80}Pt_{20}$ amorphous alloy has been investigated by anomalous X-ray scattering coupled with reverse Monte-Carlo simulation. Pt prefers to construct a Zr-rich icosahedral complex with unique chemical and topological short-range ordering. The present analysis also revealed the medium-range ordering of Pt-Pt pairs corresponding to the pre-peak signal at 17 nm^{-1} in the diffraction profile. The chemical and topological features around Pt and bonding structure correspond well with those of Mn in the approximant phase of α-AlMnSi.

1. Introduction

The short- to medium-range atomic ordering in amorphous alloys has attracted considerable attention because the interesting properties of amorphous alloys are closely associated with their atomic arrangement. Previous structural reports of Zr-based amorphous alloys have introduced an icosahedral-like atomic arrangement. It has been recognized that this structural feature improves the stability of the super-cooled liquid state and also promotes the formation of nano-sized quasicrystals [1]. The study of Zr-Pt and Zr-Pd amorphous alloys suggested that the formation of the icosahedral phase was attributed to possible icosahedral medium-range ordering (MRO) [2]. In particular, further X-ray absorption fine structure (XAFS) analysis of $Zr_{80}Pt_{20}$ amorphous alloys led to an interesting suggestion that the unique local structure around Pt is a key to elucidating the properties of the $Zr_{80}Pt_{20}$ amorphous alloy [3]. Simultaneously, the Pt-rich Zr-Pt amorphous alloy exhibited a pre-peak signal at about $Q = \sim 17\ nm^{-1}$, and the pre-peak was suggested to be related to the medium-range order comprised by Pt-centered clusters [4, 5]. In these contexts, it was concluded that the $Zr_{80}Pt_{20}$ amorphous structure comprises icosahedral MRO related to the Pt-centered cluster. This prompts us to apply the state of the art anomalous X-ray scattering (AXS) technique coupled with reverse Monte Carlo (RMC) simulation [6] for the analysis of the $Zr_{80}Pt_{20}$ amorphous alloy. AXS analysis utilizing the anomalous dispersion effect near the absorption edge provides environmental structural information about Zr and Pt. That analysis, combined with RMC simulation, allows us to discuss the three-dimensional (3D) atomic configuration of the amorphous structure.

2. Experimentals

$Zr_{80}Pt_{20}$ amorphous ribbons of approximately 20 μm thickness and 2 mm width were produced by the single roller melt-spinning technique in an Ar atmosphere. The density of the amorphous ribbon sample, measured by Archimedes' method, was 8.74 Mg/m^3. AXS measurements at the Pt L_3 absorption edge were carried out at the beam line station 7C, Institute of Materials Structure Science, High Energy Accelerator Research Organization, Tsukuba, Japan. AXS measurements at the Zr *K* absorption edge were carried out at the beam line station NW10A. Each pair of incident energies, corresponding to 25 eV and 300 eV below the absorption edge, was used in the present AXS measurements, and the anomalous dispersion terms for Pt and Zr were calculated using the relativistic Cromer and Liberman method [7]. After correcting for absorption and fluorescent radiation, the observed intensity was subsequently converted to electron units per atom using the generalized Krogh–Moe–Norman method. The Compton scattering effect was considered by using the theoretical values for a free atom with the so-called Breit–Dirac recoil factor. For the RMC simulation, we started with an initial model of 20,000 atoms (Zr: 16000 and Pt: 4000) with a dense random packing (DRP) structure in a cubic hyper-cell with a unit size of $L = 8.102$ nm.

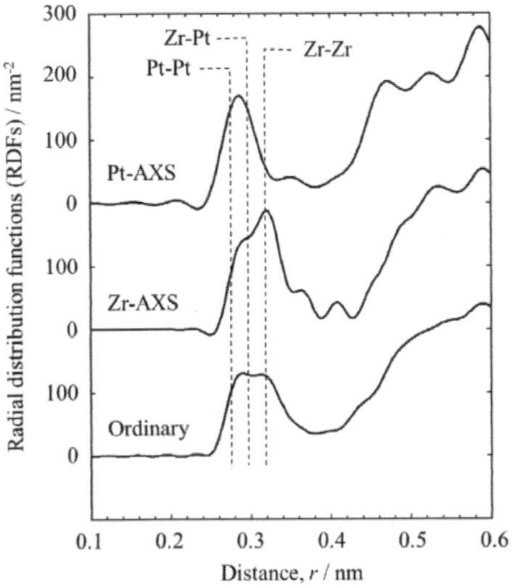

Figure 1. Ordinary- and environmental- RDFs around Pt and Zr of $Zr_{80}Pt_{20}$. Ordinary RDF was obtained from the intensity profile measured at 17.699 keV.

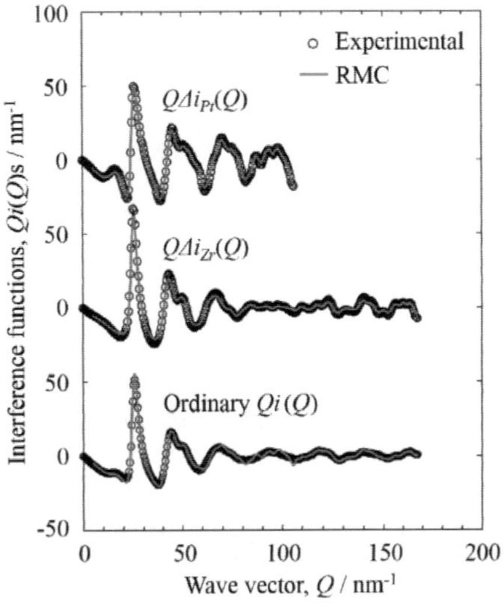

Figure 2. Ordinary- and environmental-interference functions. The solid line and dots correspond to the experimental and AXS-RMC simulation, respectively.

3. Results and Discussion

Figure 1 shows the ordinary RDF and environmental RDFs around Zr and Pt obtained by AXS measurements. The vertical dashed lines indicate the nearest neighbor atomic distances for Pt-Pt, Zr-Pt, and Zr-Zr pairs (Pt-Pt: 0.276 nm, Zr-Pt: 0.298 nm, Zr-Zr: 0.320 nm) estimated from the Goldschmidt radii. The pair correlation corresponding to the nearest Zr-Zr pair around 0.320 nm disappears in the environmental RDF around Pt. The environmental RDF around Zr indicates two types of pair correlations of Zr-Pt and Zr-Zr pairs. The observed first-nearest Zr-Pt distance is in the intermediate region between the distances for Pt-Pt and Zr-Pt pairs estimated from the Goldschmidt radii. This short Zr-Pt inter-atomic distance corresponds well with the small mixing enthalpy between Zr and Pt (−100 kJ/mol) [8]. It is interesting to note some features found in the experimental interference functions in Figure 2. A pre-peak signal is observed at approximately $Q = 17$ nm^{-1} in the functions of $Qi(Q)$ and $Q\Delta i_{Pt}(Q)$, but not in the function of $Q\Delta i_{Zr}(Q)$. This result also suggests that the pre-peak originates from the Pt-Pt interaction in the medium-range correlation. The excellent agreement between the experimentally obtained $Q\Delta i_{Pt}(Q)$, $Q\Delta i_{Zr}(Q)$, and $Qi(Q)$ and that of the RMC structural model, including the pre-peak signal, indicates that the structural model is extremely useful for the discussion of elemental selectivity and MRO. The structural parameters of averaged atomic distance and coordination number as well as other key parameters in the nearest neighbor region are summarized in Table 1. The structural information for the Zr-Pt pair suggests that there is high chemical affinity between Pt and Zr in the first-nearest Pt-Zr pair. As an example, the short distance of p_{PtZr} in comparison with that estimated from the Goldschmidt radii, clearly suggest the chemically strong bond associated with the small mixing enthalpy between Zr and Pt. In order to evaluate the topological features in the first-neighbor region, Voronoi polyhedral analysis was carried out.

Table 1 The short range ordering found in amorphous $Zr_{80}Pt_{20}$ alloy. (The averaged atomic distance, r_{ij}, the first peak top position p_{ij}, the averaged coordination number Z_{ij}, and chemical occupancy in the nearest neighbor region.)

i	j	r_{ij} [nm]	p_{ij} [nm]	Z_{ij}	Occup.(%)
Zr	Zr	0.327	0.319	11.3	82.5
	Pt	0.311	0.282	2.4	17.5
				13.7	
Pt	Zr	0.311	0.282	9.5	86.4
	Pt	0.284	0.270	1.5	13.6
				11.0	

Figure 3 shows the frequency distribution of typical Voronoi indices of Pt observed in the $Zr_{80}Pt_{20}$ amorphous alloy. The topological features around Zr are rather featureless and reproduced well by the DRP model. On the other hand, Pt has a strong preference for short-range ordering (SRO) in an icosahedron-like fashion, (0 2 8 2 0) and (0 2 8 1 0). This result allows us to consider the present Pt-centered icosahedral SRO as one of the most important structural units driven by the strong chemical affinity of the Pt-Zr pairs. The present structural analysis of the SRO in the first-nearest-neighbor region is useful for understanding the fundamental structural features in amorphous alloys; however, it is sometimes necessary to extend the model to the second-nearest-neighbor region. The present case of the $Zr_{80}Pt_{20}$ amorphous alloy could be included in this category because the pre-peak signal associated with the enhanced correlation at approximately 0.45 nm is due to the Pt-Pt correlation. The common-neighbor analysis (CNA) [9] is useful for this purpose, and the present paper demonstrates the results of the CNA for the MRO of such a Pt-Pt correlation. Figure 4 shows the results of CNA for the Pt-Pt root pairs in the second peak of the partial $g_{PtPt}(r)$. The root-pairs in this region were ranked mainly into three CNA indices of $[333]_{CN}$, $[322]_{CN}$, and $[211]_{CN}$ and the three connections indicate the remarkably higher Zr population in the common neighbors. This result readily suggests that the second-nearest Pt-Pt pairs closely associated with the pre-peak signal accommodate common neighbors of Zr rather than Pt so as to produce Zr-rich chemical short-range ordering around Pt. It may be interesting to compare the structural features of $Zr_{80}Pt_{20}$ with those of the α-AlMnSi approximant with a typical Mackay cluster [10]. Voronoi polydehdral analysis around Mn in α-AlMnSi indicates the preference for the icosahedral-like arrangement of Al with Voronoi indices of (0 0 12 0 0) and (0 2 8 2 0). At the same time, connections with the indices $[333]_{CN}$ is also common for MRO for the Mn-Mn root pairs. The similarity between the structure around Pt in $Zr_{80}Pt_{20}$ amorphous alloy and that around Mn in the quasicrystalline approximant α-AlMnSi suggests that their structures are well associated with that of the icosahedral quasicrystal.

4. Conclusion

The structure of $Zr_{80}Pt_{20}$ amorphous alloys was determined using the AXS-RMC method, and the obtained 3D structural model accurately reproduced the ordinary and environmental interference functions. Voronoi analysis indicated the overall preference for an icosahedral and icosahedral-like local structure around Pt, together with the strong chemical affinity of the Pt-Zr pair. CNA demonstrates the unique MRO of Pt-Pt, which is attributed to the connection of the SRO around Pt. The structural features around Pt are quite similar to those around Mn in α-AlMnSi. This similarity around Pt is suggested to be one of the factors responsible for the formation of the icosahedral phase during low temperature annealing.

Acknowledgments

The present AXS experiment was performed with the approval of the Photon Factory Advisory Committee (Proposal No. 2014G665). This work was financially supported in part by a Grant-in-Aid for Scientific Research (B) (16H04210) from the Japanese Society for the Promotion of Science (JSPS).

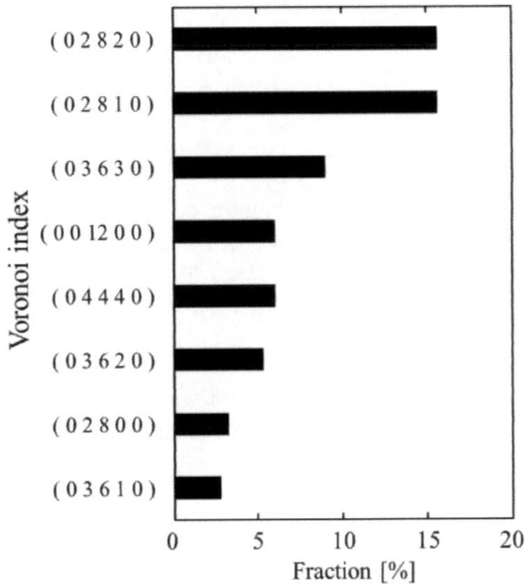

Figure 3. Frequency distribution of the typical Voronoi indices of Pt.

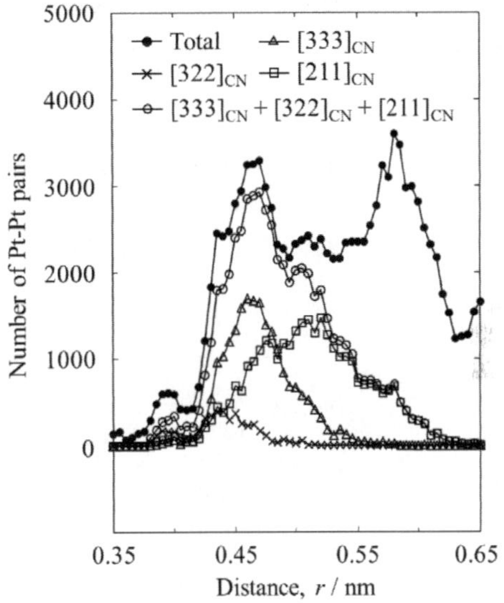

Figure 4. Results of CNA for the Pt-Pt root pairs in the MRO.

References

[1] Saida J, Imafuku M, Sato S, Matsubara E and Inoue A 2007 *J Non-Crsyt. Solids* **353** 3704.
[2] Saida J Matsushita M and Inoue A 2002 *J. Alloys Compd.* **342** 18.
[3] Saida J, Sunada T, Sato S, Imafuku M and Inoue A 2007 *Appl. Phys. Lett.* **91** 111901.
[4] Nakamura T, Matsubara E, Sakurai M, Kasai M, Inoue A and Waseda Y 2002 *J. Non-Cryst. Solids.* **312-314** 517.
[5] Sordelet D J, Ott R T, Wand S Y, Wang C Z, Besser M F, Liu A C Y and Kramer M J 2008 *Met. Mater. Trans.* **39A** 1908.
[6] Kawamata T, Yokoyama Y, Saito M, Sugiyama K and Waseda Y 2010 *Mater. Trans. Japan Inst. Metals* **51** 1796.
[7] Cromer D T and Liberman D 1970 *J. Chem Phys.* **53** 1891.
[8] Takeuchi A and Inoue A 2005 *Mater. Trans. Japan Inst. Metals* **46** 2817.
[9] Honeycutt J D and Anderson C 1987 J. Phys. Chem **91** 4950.
[10] Sugiyama K, Kaji N and Hiraga K 1998 Acta Crystallogr. **C54** 445.

Foundations of the quasicrystalline structure description in the statistical approach

Janusz Wolny, Ireneusz Buganski, Maciej Chodyn, Radoslaw Strzalka

Faculty of Physics and Applied Computer Science, AGH University of Science and Technology, Krakow, Poland

wolny@fis.agh.edu.pl

Abstract. The statistical approach was in constant development since it was first applied to calculate the structure factor for Fibonacci chain being the model of 1D quasicrystal. Its pertinence was investigated throughout years resulting in fully developed mathematical formalism finally ready to be exploited in its full potential. First serving as an alternative to higher-dimensional description of quasicrystals it proved to be even more general. Structures without *nD* representation can be investigated within statistical approach as well. Unified description of the variety of structures is possible based on the distribution function constructed upon the reference lattice concept. Furthermore, the distribution is sensitive to disturbances in the structure and reflects them by deformation occurring in the shape of the distribution. Therefore, disorder such as phononic or even phasonic, out of reach for *nD* approach, can be represented and incorporated in the structural analysis. In the paper we briefly introduce the concept standing behind the statistical approach and discuss the most important achievements of the approach.

1. Introduction – foundations of the statistical approach

The essential mathematical object considered in the statistical approach is the distribution function constructed for the structure of periodic or aperiodic crystals [1, 2]. It is the distribution of atomic positions calculated against nodes of the periodic reference lattice. The reference lattice constant λ_k is related to the length k of the scattering vector **k** according to equation $\lambda_k = 2\pi k^{-1}$. The scattering vector **k** can be arbitrarily selected from the diffraction pattern of the structure and, in fact, does not need to be the position of the Bragg peak. The peak profile can be probed within this method as well [3]. The distribution constructed for scattering vector **k** will allow to calculate the structure factor for all scattering vectors equal to n**k**, where $n \in \mathbb{Z}$. The coordinate u in the reference lattice of the jth atom is denoted as u_j and is calculated using the coordinate x_j:

$$u_j = x_j \bmod \lambda_k. \tag{1}$$

The coordinate x_j is calculated in the direct space along the direction given by vector **k**. Equation (1) simply says that each new coordinate is achieved by projection of the atomic coordinate onto reference lattice and the distance to the nearest node of the lattice becomes new coordinate. The above mentioned procedure is mathematically realized by modulo operation. All coordinates u_j are limited to the region $[-\lambda_k/2, \lambda_k/2)$ or equivalently $[0, \lambda_k)$ depending on the taken definition. Both regions are equally correct in terms of the structure factor calculation. The difference is reflected in the global phase, which is insignificant. Having obtained coordinates u we can construct the distribution $P(u)$, which gains probabilistic interpretation after normalization. If the considered crystal structure exhibits multiple characteristic length-scales, e.g. in the case of modulated structures, the construction of the distribution function can be generalized. In case of two length-scales-structure each scattering vector $\mathbf{k}$ can be expanded as the sum of the basic vector $\mathbf{k}_0$ and the modulation vector $\mathbf{q}_0$: $\mathbf{k} = n\mathbf{k_0} + m\mathbf{q_0}$, $n, m \in \mathbb{Z}$. The length q_0 of the modulation vector is related to the length k_0 of the basic vector by scaling relation, such that scaling $\xi = k_0/q_0$. By analogy, we introduce reference lattice with lattice constant λ_q related to the length of the modulation vector in the same manner, as in case of the basic vector. Coordinates respective to the second lattice are denoted as v and are obtained in the same procedure as u coordinates. Coordinates v are limited to $[-\lambda_q/2,\ \lambda_q/2)$ or $[0,\ \lambda_q)$. The complete information about the structure with two length-scales is stored in the two-parameter-distribution $P(u, v)$. Generalization of the statistical distribution for rank *n* quasicrystal would require *n* parameter-dependent-distribution but it must be emphasized that such generalization is not necessary and the distribution dependent on one parameter u is sufficient to calculate the structure factor for the scattering vector **k**. The distribution $P(u, v)$ is a fundamental definition of Average Unit Cell (AUC) but in most applications, as it was mentioned, it is enough to use marginal distribution only. It is straight-forwardly related to the scaling property of the basic vector and the modulation vector. The scaling relation is also present in mutual dependence of u and v. The existence of relation $v(u)$ imposes the possible simplification of the probabilistic structure description. The distribution is always non-zero only in region realized by $v(u)$ function which is a set of curves or single curve in simple case. What we refer to as AUC is a marginal distribution $P(u)$. A full diffraction pattern can be calculated form $P(u)$ distribution without its general counterpart. However, if multi-scaling description of the AUC is used such simplification is valid for non-disordered structures only as long as relation $v(u)$ is preserved. Phononic disorder affects the scaling, therefore full distribution is required [4]. The structure factor for 1D structure with two length-scales in kinematical theory is the sum of plain waves scattered over atoms in the structure. It can be written in statistical formalism as:

$$F(nk_0 + mq_0) = \sum_{j=1}^{N} f_j \exp\left[ink_0u_j + imq_0v_j\right], \tag{2}$$

where scattering vector $k = nk_0 + mq_0$, $n, m \in \mathbb{Z}$ and f_j is the atomic form factor. We can use lengths of vectors because coordinates u and v are calculated along those vectors. Taking into account that coordinates u_j and v_j form dense distribution $P(u, v)$ the formula (2) can be rewritten as:

$$F(nk_0 + mq_0) = \int_{\mathrm{AUC}} P(u, v)\exp[ink_0u + imq_0v]\,\mathrm{d}u\mathrm{d}v. \tag{3}$$

Further simplification is possible due to scaling relation $v(u)$:

$$\begin{aligned} F(nk_0 + mq_0) &= \int_{\mathrm{AUC}} P(u, v)\exp[ink_0u + imq_0v(u)]\,\mathrm{d}u\mathrm{d}v \\ &= \int_{\mathrm{AUC}} P(u)\exp[ink_0u + imq_0v(u)]\mathrm{d}u. \end{aligned} \tag{4}$$

Formula (4) is general for any system with scaling relation and two length-scales. Even further simplification can be achieved when specific form of $v(u)$ relation is applied, e.g. linear function in case of ordinary quasicrystals. The most popular scaling for known quasicrystals is τ – the golden mean ratio. Respective scaling relation $v(u)$ is linear with coefficient equal to $-\tau^2$.

The second most important object in the statistical approach after distribution function together with $v(u)$ relation is the envelope function [2]. The envelope function is calculated as the Fourier Transform of the statistical distribution. Its applicability is particularly important in the case of quasicrystals. The diffraction pattern of quasicrystals is aperiodic but peaks can be grouped into sets. Peaks with the same m value can by connected with a single curve called envelope (Figure 1). In general, the shape of the envelope function can be complex. The curves occur periodically in the reciprocal space with the periodicity k_1 depending on the scaling. The set of envelope functions form the equivalence class and all peaks can be represented by only one envelope function in the space of the so-called reduced vector $w = k - mk_1$ [2]. It is very convenient to use the envelope function representation of the diffraction pattern as numerous effects influence the shape of the envelope function. For instance, phasons are easily recognizable by reshaping the envelope function [5].

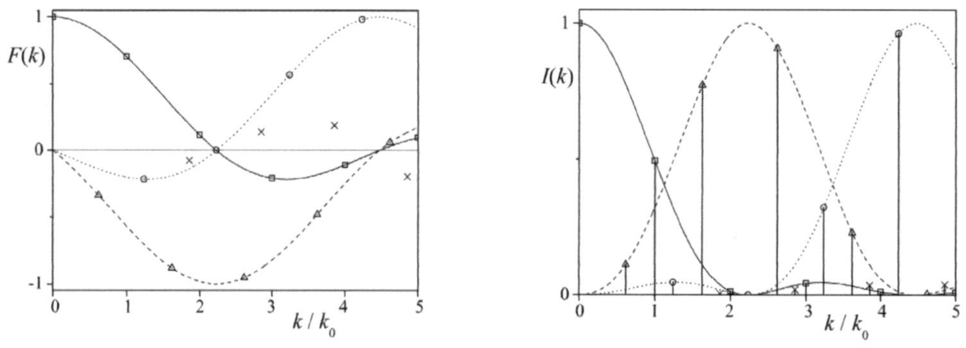

Figure 1. (left) The structure factor $F(k)$ calculated for the Fibonacci chain using formula (4) with first 3 envelopes $m = \{0,1,2\}$ marked with different line types (solid, dashed, dotted respectively). Different marker types were used to distinguish peaks with respect to m value; (right) The diffraction pattern of the Fibonacci chain calculated as squared structure factor with first 3 envelopes marked (notation from the left figure used). Peaks are distributed periodically within an envelope (period $k_0 = \frac{2\pi\tau^2}{\tau^2+1} \approx 4.54\,[a.u.]$) and envelopes are deployed with periodicity $k_1 = \sqrt{5}k_0 \approx 10.2\,[a.u.]$.

2. The refinement of quasicrystals

The refinement of quasicrystalline structure is done based on the rigid body model where we take the advantage of aperiodic tilings. Also the cluster description of atomic arrangement is possible and the statistical model could be derived for clusters as well [6]. Tilings have this advantage over clusters that neighboring tiles do not overlap and number of prototiles is rather low. In the tiling models we assume identical decoration for each tile of the same type independently on the orientation. In general form, the geometric structure factor for tiling model can be written as follows:

$$F(\mathbf{k}) = \sum_{i=1}^{N_p} \sum_{j=1}^{N_o} F(\mathbf{k})_{p,o} \sum_{l=1}^{N(p)} f_l \alpha_l \exp(i\mathbf{k} \cdot \mathbf{r}_{p,o,l}), \tag{5}$$

where $F(\mathbf{k})_{p,o}$ is the Fourier Transform of this part of the distribution $P(u)$ which is related to the reference node of the prototile p. Number of prototiles is equal to N_p. Each prototile is in the orientation o and total number of orientations is N_o. Atomic decoration is considered in the last sum, where f_l is the scattering factor of lth atom and α_l describes the fraction of atom inside a prototile as atoms can be shared by more prototiles. Number N of atoms in a prototile is dependent on the type of prototile, what is denoted as $N(p)$. The position of atom l inside the prototile p in the orientation o is given by $\mathbf{r}_{p,o,l}$. The key part of the calculation is to determine the distribution of prototile's reference vertex in the reference lattice in the purpose of reducing the complexity of calculations. The complete structure factor was derived for decagonal quasicrystals modeled with the use of Penrose Tiling (PT) [7] and icosahedral quasicrystals modeled with Ammann-Kramer-Neri Tiling known as Ammann Tiling (AT) [8]. In further parts of the paper prototiles are called structure units.

2.1 The Penrose Tiling

The Penrose Tiling covers 2D space with the use of two structure units: thick and thin rhombuses [9]. Each rhombus is spanned by two vectors associated with pentagonal direct space vector basis and occurs in five different orientations. If the AUC is constructed according to procedure given in section 1 the distribution for vertex decoration PT is uniform and non-zero only in four pentagonal regions [7]. It is not a coincidence that atomic surface of the PT is also composed of four pentagons. In fact, AUC in case of quasicrystals can be considered as oblique projection of atomic surface onto the physical space [7]. In order to express each atomic position in relation to one reference vertex of the rhombus its distribution in the AUC must be found. The Fourier Transform of such created subregions of the distribution is equal to $F(\mathbf{k})_{p,o}$ in the formula (5). The statistical method has been already used to refine the structures of decagonal quasicrystals e.g. AlCuMe, Me={Co, Rh, Ir} alloys [10]. Further refinement can be done by using Generalized Penrose Tiling (GPT) instead of standard PT [11]. In higher-dimensional picture GPT can be generated by shifting higher-dimensional projection strip by a fraction of unit vector, resulting in change of shape of atomic surfaces. Three of previously pentagonal atomic surfaces become decagons and one additional pentagonal atomic surface is created. Projecting higher dimensional periodic lattice through such atomic surfaces generates GPT. The atomic structure depends on the shift. The unit tiles of GPT are the same two Penrose rhombuses as in PT, however the matching rules are different. GPT does not belong to Penrose mutual local derivability (PMLD) class of tilings, so each of the tilings will have slightly different diffraction pattern (example of how diffraction peak intensity changes in function of shift s is presented in Figure 2 left). This allows to use shift as additional refinement parameter during structure refinement, effectively allowing us to fit long range order of the structure, resulting in the improvement of a model [11] – see Figure 2 (right).

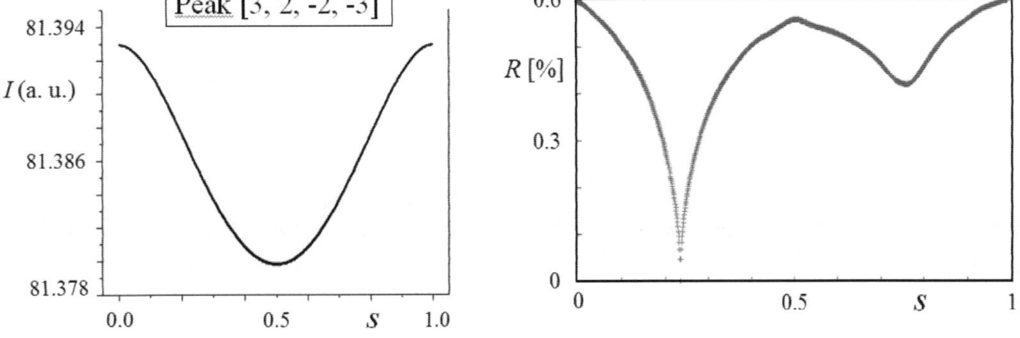

Figure 2. (left) Change of diffraction peak's relative intensity in function of shift s; (right) R-factor in the function of shift, calculated for the GPT with simple decoration (vertex decoration and three identical atoms per rhombus located inside). Different depths of minima correspond to the lack of symmetry of atomic decoration inside building units. In case of symmetrical units (e.g.. atoms only in vertices of building units or no decoration) those minima are the same. Parameter $s=1$ means that we return to the initial position with four pentagons.

2.2 The Ammann Tiling

3D quasicrystals are known to exhibit icosahedral point symmetry in the reciprocal space. The structure of icosahedral quasicrystals can be modeled with the use of Ammann Tiling [12]. It uses two golden rhombohedra: obtuse (oblate) and acute (prolate) to cover 3D space. The statistical distribution constructed for vertex decoration of the AT is also uniform as in the case of PT and non-zero only in region limited to rhombic triacontahedron [8]. It was mentioned before that AUC is the oblique projection of the atomic surface onto the physical space, therefore similarity between the shape of atomic surface and AUC for AT is obvious. The subdivision of the AUC must be performed with respect to the reference node of each of two rhombohedra. Obtained subregion has the shape of obtuse rhombohedron in the case of obtuse structure unit and acute rhombohedron for acute structure unit (Figure 3). There are ten possible orientations for each rhombohedron.

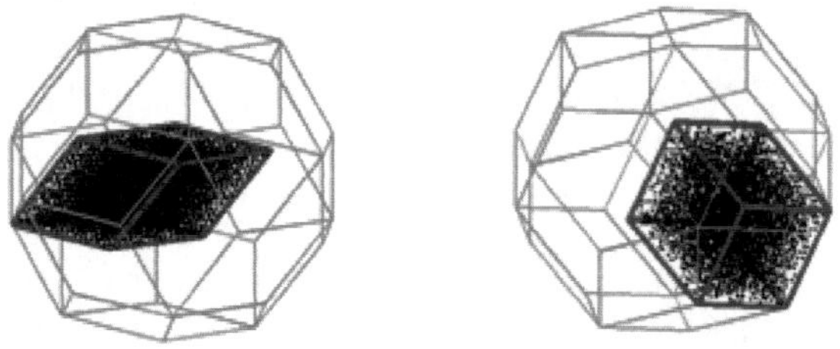

Figure 3. Subregions belonging to the reference vertex of acute rhombohedron (left) and obtuse rhombohedron (right) inside the region of non-zero statistical distribution (red rhombic triacontahedron).

3. Corrections to the structure factor

The main purpose of the structure refinement is to obtain the best structural model with respect to experimental data. Structural information can be obtained from the X-ray diffraction pattern. The calculation of structure factor allows to compare theoretical model of the structure with peaks' intensities collected in the diffraction experiment. Despite geometric structure factor we need to consider effects changing peak intensity including phononic and phasonic disorder [13]. The statistical method allows to incorporate both effects within one uniform formalism. If we assume independent atom fluctuation from the equilibrium position [14] by the distance subjected to any kind of distribution the statistical distribution $P(u,v)$ will be smeared along (1,1) direction [3] – see Figure 4. In case of quasicrystals with linear scaling relation smearing leads to a broadening of the $v(u)$ line. To properly calculate the structure factor we need to consider full $P(u,v)$ distribution, but calculations lead to a simple result. Structure factor is the product of geometric structure factor and the correction term. If atomic positions are subjected to Gaussian distribution the correction takes the form of Debye-Waller factor [15]. The phononic correction for quasicrystals is the same as for periodic crystals.

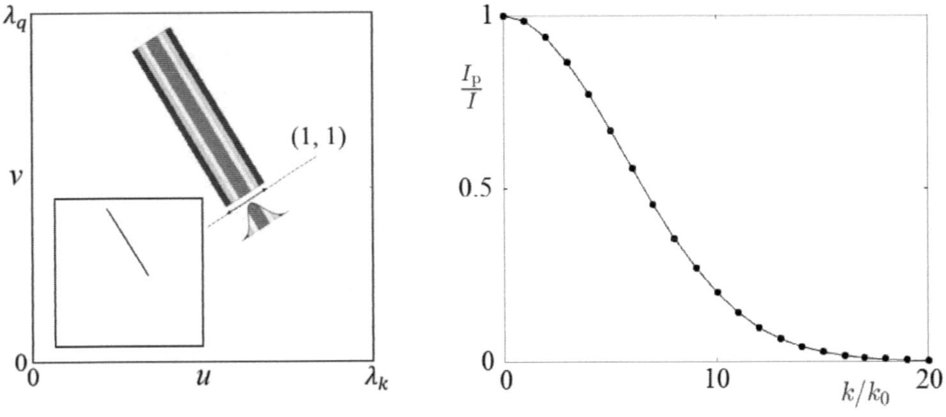

Figure 4. (left) The AUC and TAU2-scaling relation for Fibonacci chain modified by Gaussian phononic smearing. Respective $v(u)$ relation for ideal structure is shown in the inset; (right) The phonon correction term calculated for quasicrystal (solid line) and periodic crystal (full circles). The intensity of the peak for the structure with phonons (I_p) was normalized by intensity of the peak for the structure without phonons (I). In both plots the amplitude of phonons was 2% of λ_k (the standard deviation). Main scattering vector $k_0 \approx 2.81$ $[a.u.]$.

It is a well-known fact that the underestimation of the weak peaks' intensities is observed in the calculated vs. observed intensities plots for quasicrystals. Recently, the observation was made with a conclusion that improper handling of phasonic disorder can lead to biased intensities of reflections calculated within a model [5]. In particular, generalized Debye-Waller correction may be responsible for the above mentioned effect (Figure 4 - right). Based on the statistical method we are able to handle the problem of phasons analytically by taking into account geometric restrictions of phason flips. Improvement of resulted model in terms of R-factor is for now proven in computer simulations [2].

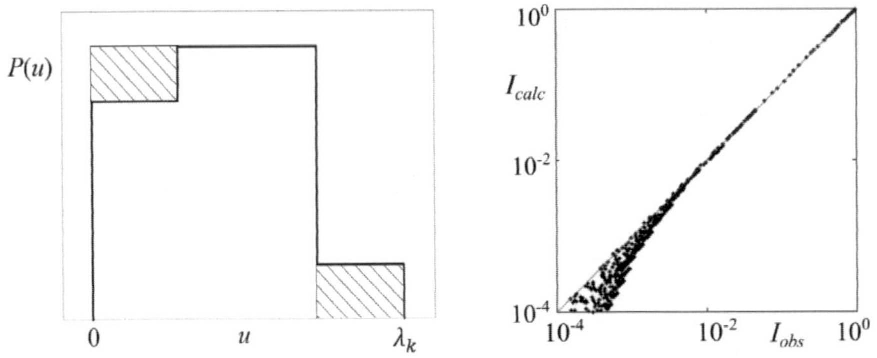

Figure 4. (left) The AUC for the Fibonacci chain with phasons (flip ratio $\alpha = 20\%$); (right) Calculated *vs.* "observed" intensities in the diffraction diagram of Fibonacci chain with phasons ($\alpha = 5\%$) and standard phasonic Debye-Waller correction. The characteristic bias is observed for weak reflections ($I < 10^{-2}$).

Conclusions

The statistical approach is a complete mathematical theory serving description of periodic and non-periodic structures. Its foundation is based on the possibility of constructing distribution function with full structural information necessary to calculate the structure factor. Although the method can be applied to periodic crystals as well, its full potential is seen as far as aperiodic systems are considered. In this case, traditional unit cell is being replaced by the Average Unit Cell. The advantage of the method is seen in its applicability. It can be constructed for structures with no *nD* representation and therefore can be used beyond limitations of higher-dimensional approach. The statistical approach allows to incorporate both phononic and phasonic disorder. Although phononic correction can be considered in Debye-Waller form for quasicrystals as well, its phasonic generalization seems to fail in weak-peaks-regime. Based on statistical formalism we are able to handle phasons in fully analytical manner. As a result, more accurate structure models are achievable.

Acknowledgments
Authors kindly acknowledge financial support from Polish National Science Center under Grants No. DEC-2013/11/B/ST3/03787 (all) and 2014/13/N/ST3/03776 (R.S.). I.B. and M.C. kindly acknowledge financial support from Marian Smoluchowski Krakow Research Consortium "Matter Energy-Future" under grant KNOW.

References
[1] Wolny J Kuczera P Strzalka R 2015 *Appl. Phys. Lett.* **106**, 131905
[2] Wolny J Kozakowski B Kuczera P Pytlik L and Strzalka R 2014 *Acta. Cryst.* A. **70** 181-5
[3] Strzalka R Buganski I and Wolny J 2016 *Crystals* **6** 104
[4] Wolny J Buganski I Kuczera P and Strzalka R 2016 *J. Appl. Crystallogr.* Submitted
[5] Buganski I Strzalka R and Wolny J 2016 *Phys. Status Solidi* B. **253** 450-7
[6] Wolny J Kozakowski B and Dabrowska A 2006 *Philos. Mag.* **86** 637-43
[7] Kozakowski B and Wolny J 2010 *Acta Crystallogr.* A. **66** 489-98
[8] Strzalka R Buganski I and Wolny J 2015 *Acta Crystallogr.* A. **71** 279-90
[9] Steurer W and Deloudi S 2009 *Crystallography of Quasicrystals:Concepts, Methods and Structures* (Heidelberg: Springer-Verlag Berlin) pp 34-42
[10] Kuczera P Wolny J and Steurer W 2012 *Acta Crystallogr.* B. **68** 578-89
[11] Chodyn M Kuczera P and Wolny J 2015 *Acta Crystallogr.* A. **71** 161-8
[12] Kramer P and Neri R 1984 *Acta Crystallogr.* A. **40** 580-7
[13] De Boissieu M 2012 *Chem. Soc. Rev.* **41** 6778-86
[14] Burgi H-B 1995 *Acta Crystallogr.* B. **51** 571-9
[15] Henley C Elser V Mihalkovic M 2001 *Z. Kristallogr.* **215** 553-68

Atomic structure of the primitive cubic phase P_{40} in the Al-Pd-Ru system

Y Hatakeyama[1], **N Fujita**[1,2] **and A P Tsai**[1,3]

[1] Institute of Multidisciplinary Research for Advanced Materials, Tohoku University, Sendai 980-8577, Japan
[2] JST, PRESTO, Kawaguchi, Saitama 332-0012, Japan
[3] National Institute for Materials Science, Tsukuba 305-0047, Japan

E-mail: nobuhisa@tagen.tohoku.ac.jp

Abstract. The atomic structure of the high-order cubic approximant P_{40} in the Al-Pd-Ru system is determined by means of single-crystal X-ray diffraction.

1. Introduction

Icosahedral quasicrystals (iQC's) in Al-Pd-TM systems, where TM stands for transition metal elements (e.g., Mn, Re, Ru, Os), are among the oldest families of thermodynamically stable F-type iQC's [1, 2]. These quasicrystals continue to offer attractive challenges to unravel their atomic structures because of a number of uncertainties in their fine structural details. Albeit the study of approximant phases generally provides information that can help us understand local atomic arrangements in the iQC's, suitable approximants have seldom been reported in a compositional proximity to the iQC's. In the relevant ternary phase diagrams, there often exist one or a few cubic phases (C, C_1 and/or C_2 phase) known as the 1/0 cubic approximants lying next to the iQC's. Still, the stability ranges of the C-phases are somewhat detached from those of the iQC's, and besides their unit cells may be too small to abstract many of the complex features of the iQC's.

Perhaps, more insights could be given by higher order approximants, which for instance have been reported in ternary Al-Pd-Co [3] and Al-Pd-Re [4] systems and also in quaternary Al-Pd-Mn-Si [5, 6], Al-Pd-Cr-Fe [7] and Al-Pd-Ge-Co [8] systems. These authors have derived a few kinds of atomic cluster which may potentially constitute the stable iQC's in Al-Pd-TM systems. However, the things may not work out as we expect because views on the constituent clusters are not necessarily consistent among these papers. Another concern is that any of these approximants does not accompany a stable iQC in its immediate proximity of composition. For elucidating the structural relationship between the iQC's and approximants, a sensible approach would be to focus on an alloy system in which both an iQC and a high-order approximant lie in close proximity to each other in the phase diagram.

In the Al-Pd-Ru alloy system, the stability range of the iQC roughly extends between the compositions $Al_{73}Pd_{11.5}Ru_{15.5}$ and $Al_{70}Pd_{22}Ru_8$ at a temperature between 790°C and 900°C [9]. At the high-Ru end of this range lies a narrow region of a large cubic approximant, P_{40}, with $Pa\bar{3}$ space group and a lattice constant of about 40Å[9]. At the low-Ru end on the other hand lies a narrow region of another approximant, P_{20}, which is often called the 2/1 cubic approximant,

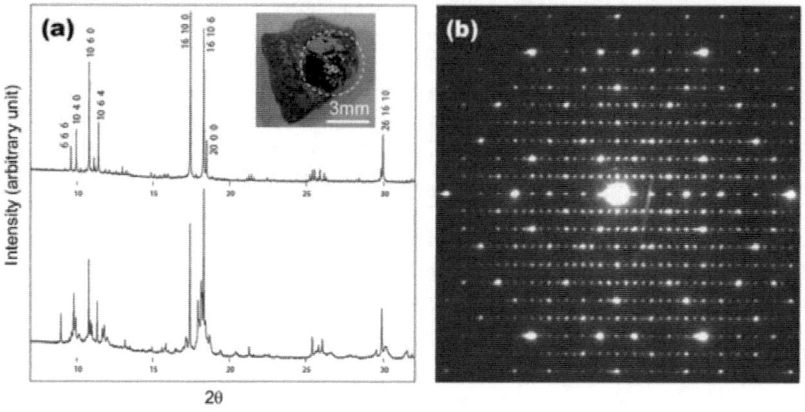

Figure 1. (a) Powder diffraction patterns with synchrotron X-ray (wave length=0.652802 Å) taken from the single-crystal part (upper panel) and polycrystalline part (lower panel). The inset is a photo image of the sample, in which the single-crystal part is indicated with a dotted circle. (b) A selected area electron diffraction pattern taken from the single-crystal part with the ⟨100⟩ incidence, indicating the formation of the P_{40} phase.

with a lattice constant of about 20Å[10, 11]. Because of the compositional proximity between the iQC and the two approximants, the present system offers a rare opportunity to investigate the structural relationship between them. So far, however, the structural details of both the P_{40} and P_{20} phases have yet to be clarified.

In this study, the crystal structure of P_{40} is elucidated for the first time using single-crystal X-ray diffraction. The aim of this paper is to provide a short account of the structure analysis, while more details will be presented in a separate paper. In fact, remarkable similarities can be recognized between published electron diffraction patterns for P_{40} [9] and ones for the Al-Pd-Cr-Fe approximant [7]. Our present results ensure that the crystal structures of the two approximants are identical.

2. Synthesis

An ingot with the composition $Al_{72.6}Pd_{13}Ru_{14.4}$, which has been reported to be the one for P_{40}[9], was prepared by ordinary arc-melting under an argon atmosphere. Three parts of the ingot were annealed at 900 °C, 1000 °C and 1100 °C for a few days. All the samples were found to contain multiple phases including the P_{40} and C_2 phases, while powder X-ray diffraction indicated that the sample annealed at 1000 °C had the largest amount of P_{40}. Hence, another ingot was prepared with the same composition and annealed at 1000 °C for 20 days. As shown in the inset of Fig. 1(a), a part of the sample revealed a single-crystal grain 3 mm in diameter wrapped in a polycrystalline matrix. Using powder X-ray diffraction Fig. 1(a) and selected area electron diffraction Fig. 1(b), the single crystal was confirmed to be that of P_{40}[9], while several kinds of crystallites are in the polycrystalline part. It seems that the single crystal has grown in a melt of residual components, which have turned into fine crystallites upon cooling. Inductively coupled plasma atomic emission spectrometry (ICP-AES) was used to analyze the composition of the single-crystal part, indicating the composition to be $Al_{72}Pd_{16.4}Ru_{11.6}$. The deviation from the nominal composition may be too large to be explained on a mere assumption that the initial ingot was somewhat inhomogeneous. Considering that no previous report (cf. [9]) used a single crystal to analyze the composition, our result should indicate the right composition of P_{40}.

3. Measurement and analysis

The single-crystal X-ray diffraction experiments were carried out at room temperature on a fragment (roughly in the shape of a cube with edge length being slightly less than 0.1 mm) of the single-crystal, using a SMART APEX diffractometer (Bruker) mouting CCD area detector with Mo $K\alpha$ radiation and graphite monochromator (wavelength 0.71073Å). Indexing and empirical absorption correction were performed using the Bruker software package (SMART, SAINT and SADABS)[12]. The Bravais lattice was identified to be primitive cubic while the lattice constant was estimated to be 40.67Å.

The indexed diffraction intensities were analyzed using the software package JANA2006 [13]. The space-group was unambiguously determined to be $Pa\bar{3}$ (No. 205) with R_{int} (obs/all) being 8.26/21.14. There were 59904 independent reflections within a 2θ range of $1.12 \sim 40.64\,^{\circ}$, wherein 12604 were classified as observed (i.e., $I > 3\sigma(I)$). The initial model for the structure refinement was obtained using the SUPERFLIP program [14] called from the JANA2006 platform. The refinement converged reasonably well with $R(F) = 9.65$ %, $R_w(F) = 10.00$ % (σ weight) and $(\Delta\rho_{\mathrm{max}}, \Delta\rho_{\mathrm{min}}) = (7.46, -4.99)$ e/Å^3. In the final model, each unit cell contains 4824 atomic sites among which 208 are symmetrically unique, whereas the number of refined parameters is 1073. The composition calculated from the refined model is $Al_{71.40}$ $Pd_{16.68}$ $Ru_{11.92}$, which agrees well with the result of ICP-AES. Since each unit cell corresponds to 43.2 formula units (4320 atoms), the density is calculated to be 5.23 g/cm^3.

4. Crystal structure

The refined structure exhibits a dense packing of atomic clusters, which are called pseudo-Mackay clusters (pMC's) and mini-Bergman clusters (mBC's). Whereas these clusters are joined together by sharing atoms, three kinds of linkage between clusters are observed. The first two kinds are parallel to 2-fold symmetry axis with the cluster centers separated by about 7.8Å. The other kind of linkage is parallel to 3-fold axis and the separation is about 6.7Å. The former two linkages connect the same kind of cluster (pMC-pMC and mBC-mBC), whereas the third linkage connects distinct cluster types (pMC-mBC). The cluster network thus determined can be represented as a tiling composed of four kinds of polyhedra called the canonical cells [15], where the edges of the cells correspond to the inter-cluster linkages. The kind of cluster at each node is determined unambiguously by the parity of the node. The present tiling can be understood as a $2\times\ 2\times\ 2$ superstructure of the standard 3/2-cubic packing [15].

We find that the centers of the clusters are mostly occupied by Pd or Ru while the inner shells mostly by Al. It is important to note that each pMC-pMC linkage involves a weak inter-penetration of the two pMC's, such that an outer-shell atom of either pMC is enforced to be dimpled in to lie on the outer shell of the other pMC. The inner shell of each pMC is not fully occupied but is occupied by 6 to 8 atoms (mostly Al), which also belong to the outer shells of adjacent mBC's connected through pMC-mBC linkages. It turns out that the present crystal structure is identical to that of the Al-Pd-Cr-Fe approximant [7].

In fact, pMC and mBC may serve as the fundamental building units of the iQC as well as the P_{20} phase, in which the arrangement of clusters may be represented as another canonoical-cell tiling. Moreover, number of hypothetical crystal structures are derived from the same principle, which may thus act as a universal framework for describing the cluster-packing geometry in Al-Pd-TM alloys. Indeed, it can be shown that the Al-Pd-Co [3] and Al-Pd-Ge-Co approximants [8] exhibit the same kind of cluster-packing geometry. Still, one may be puzzled to find that the two Al-Pd-Mn-Si approximants, usually referred to as 1/1 and 2/1 cubic approximants, have so far been described in incompatible manners [5, 6]. Here, apart from that the reported clusters are neither pMC nor mBC, the relationship between the two approximants is left unclear. In addressing this issue, a detailed structure analysis of P_{20}, which exhibits the same crystal symmetry as the 2/1 approximant in Al-Pd-Mn-Si, could give us valuable hints.

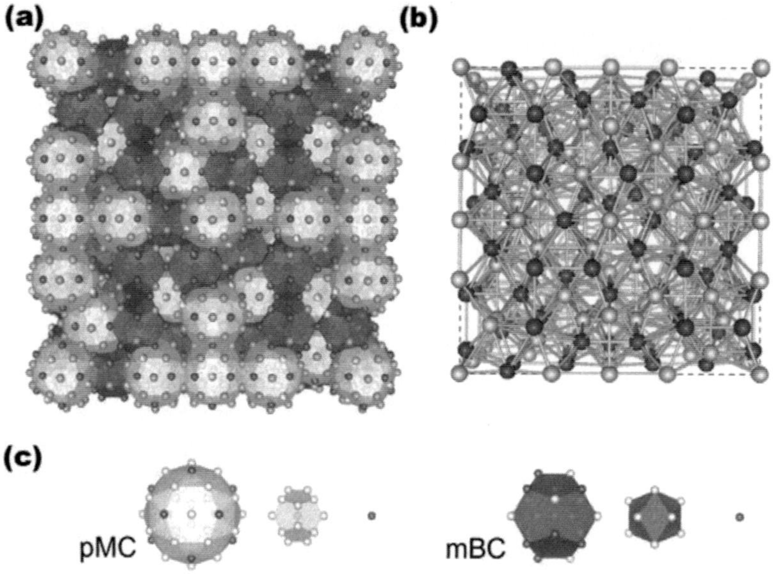

Figure 2. (a) The crystal structure of P_{40} represented as a packing of clusters. pMC's and mBC's are shown in light and dark gray, respectively. White spheres represent Al, while gray spheres Pd or Ru. (b) The cluster network extracted from panel-(a). Each sphere represents a cluster, where the two kinds of cluster are distinguished by color. Edges of the unit cell are represented as dashed lines, while cluster linkages as cylindrical bars. (c) The shell structures for pMC and mBC. Panel-(a) and panel-(b) corresponds to Figs. 7 and 9 in Ref.[7].

Acknowledgments
The authors thank Akira Sato for technical assistance in the single-crystal X-ray diffraction measurements. The synchrotron powder X-ray diffraction experiments were performed at the BL15XU beamline of SPring-8 with the approval of National Institute for Materials Science (NIMS) (Proposal No. 2015A4901). N F thanks Marek Mihalkovic for helpful discussions and comments on related subjects. Financial support from JSPS (Grant-in-Aid for Scientic Research(C) no. 25400354) and JST (PRESTO) is gratefully acknowledged (N F).

References

[1] Tsai A P, Yokoyama Y, Inoue A and Masumoto T 1990 *Jpn. J. Appl. Phys.* **29** L1161
[2] Asao T, Tamura R and Takeuchi S 2002 *Philosophical Magazine Letters* **82** 217
[3] Simura R, Suzuki S, Yubuta K and Sugiyama K 2013 *Mater. Trans.* **54** 1385
[4] Suzuki S, Simura R and Suziyama K 2011 *Philosophical Magazine* **91** 2610
[5] Sugiyama K, Kaji N and Hiraga K 1998 *Zeitschrift für Kristallographie* **213** 168
[6] Sugiyama K, Kaji N and Hiraga K 1998 *Zeitschrift für Kristallographie* **213** 90
[7] Fujita N, Takano H, Yamamoto A and Tsai A P 2013 *Acta Cryst.* A **69** 322
[8] Sugiyama K, Yubuta Y, Yokoyama Y, Suzuki S and Simura R 2014 *Acta Phys. Pol.* A **126** 588
[9] Pavlyuchkov D, Grushko B and Velikanova T Y 2009 *Journal of Alloys and Compounds* **469** 146
[10] Tamura R, Asao T, Tamura M and Takeuchi S 1999 *J. Phys.: Condens. Matter* **11** 10343
[11] Shibuya T, Asao T, Tamura M, Tamura R and Takeuchi S 2000 *Mater. Sci. Eng. A* **294-296** 61
[12] Bruker 2001 SMART (v. 5.625), SAINT (v. 6.45) and SADABS *Bruker AXS Inc., Madison, Wisconsin, USA*
[13] Patricek V, Dusek M and Palatinus L 2006 JANA2006: The Crystallographic Computing System *http://jana.fzu.cz/ (Institute of Physics, Prague, Czech Republic)*
[14] Palatinus L and Chapuis G 2007 *J. Appl. Cryst.* **40** 786
[15] Henley C L 1991 *Phys. Rev. B* **43** 993

Contribution to thermodynamic description of Al-Pd system

Libor Ďuriška[1*], Ivona Černičková[1], Roman Čička[1] and Jozef Janovec[2]

[1] Slovak University of Technology in Bratislava, Faculty of Materials Science and Technology in Trnava, Paulínska 16, 917 24 Trnava, Slovak Republic
[2] Slovak University of Technology in Bratislava, Vazovova 5, 812 43 Bratislava, Slovak Republic

*email: libor.duriska@stuba.sk

Abstract. The homogeneity range around the Al_3Pd stoichiometry, representing the family of quasicrystalline approximants denoted as ε_n and/or (Al_3Pd), was involved in the calculated Al-Pd phase diagram. The calculations were performed by means of the CALPHAD method using the Thermo-Calc software. A novel thermodynamic description of the ε_n/(Al_3Pd) phase was proposed based on the $(Al\%,Pd)_3(Al,Pd\%)_1$ two-sublattice model. Existing thermodynamic parameters of stoichiometric phases Al_4Pd and $Al_{21}Pd_8$ were slightly adjusted. The transition between phases ε_n/(Al_3Pd) and γ-$Al_{21}Pd_8$ was predicted thermodynamically at 619°C.

1. Introduction

The Al-Pd system has been studied several times so far. McAlister [1] published a provisional Al-Pd phase diagram with nine equilibrium intermediate phases in 1986. The diagram was later adopted by Massalski et al. [2]. In 2001, Yurechko et al. [3] published a revised phase diagram in the compositional range between 0 and 40 at.% Pd. The revised diagram comprises two high-temperature phases around the Al_3Pd stoichiometry. In 2003, Okamoto [4] published the phase diagram based on previous findings. The latest Al-Pd phase diagram has been published by Grushko [5], bringing new information about constitutions around both Al_3Pd and AlPd stoichiometries.

Considering previously published results, following phases have been referenced in the Al-Pd system [4,5]: (Al), λ-Al_4Pd, γ-$Al_{21}Pd_8$, ε_6 (~Al_3Pd), ε_{28} (~Al_3Pd), δ/(Al_3Pd_2), β/(AlPd), μ/(AlPd), υ/(Al_3Pd_5), ρ/($AlPd_2$), τ/(Al_2Pd_5), and (Pd). In the thermodynamic notation, all the non-stoichiometric phases are placed between round brackets in this work, e.g. β/(AlPd). Quasicrystalline approximants ε_6 and ε_{28} belong to so-called ε-family subsuming five orthorhombic phases (ε_6, ε_{16}, ε_{22}, ε_{28}, and ε_{34}) differing from each other in a lattice parameter. For the common denotation of at least two phases of the ε-family, the symbol ε_n [6] is used in this work. Although ε_6 and ε_{28} were found in Al-Pd alloys, an exact boundary between ε_6- and ε_{28}-areas has not been determined in the corresponding phase diagram [3-5].

The Al-Pd system has been thermodynamically assessed by Li et al. [7]. All the experimental phase equilibria and thermodynamic data from the available literature have been evaluated using the CALPHAD method. Despite of a good coincidence between both calculated [7] and experimental phase diagrams [5], several differences are evident around stoichiometries Al_3Pd and AlPd-Al_3Pd_5. The ε_n area, occurring in the experimental phase diagram proposed by Grushko [5], is described by stoichiometric phases Al_3Pd and $Al_{21}Pd_8$ in the calculated phase diagram [7].

The main aim of this work is to use the CALPHAD method for the calculation of the high-temperature Al_3Pd as a non-stoichiometric $\varepsilon_n/(Al_3Pd)$. Since the experimental results related to the distribution of ε_6 and ε_{28} in the ε_n area are not unambiguous yet, $\varepsilon_n/(Al_3Pd)$ was considered in thermodynamic calculations as a non-stoichiometric "single phase". At this stage, the optimization of the parameters using PARROT has not been done. Taking into account the differences between both calculated and experimental phase diagrams, the effort was paid to approach the shape of the ε_n area to that recently published by Grushko [5] by evaluation of the parameters of $\varepsilon_n/(Al_3Pd)$ in the new thermodynamic model.

2. Thermodynamic model

Li et al. [7] described Al_3Pd as a stoichiometric phase. Since the experimental results published recently [3-5] confirmed the occurrence of ε_n with homogeneity range around the Al_3Pd stoichiometry, it is thermodynamically described by the two-sublattice $(Al\%,Pd)_3(Al,Pd\%)_1$ model in this paper (the symbol % denotes the major component in the corresponding sublattice). The Gibbs energy per mole of the formula unit (Al_3Pd) is given by the following expression:

$$G_m^{(\mathrm{Al_3Pd})} - H_{(\mathrm{Al_3Pd})}^{\mathrm{SER}} = y_{\mathrm{Al}}^{(1)} y_{\mathrm{Al}}^{(2)}\, {}^0G_m^{\mathrm{Al_3Al}} + y_{\mathrm{Al}}^{(1)} y_{\mathrm{Pd}}^{(2)}\, {}^0G_m^{\mathrm{Al_3Pd}} + y_{\mathrm{Pd}}^{(1)} y_{\mathrm{Al}}^{(2)}\, {}^0G_m^{\mathrm{Pd_3Al}} + y_{\mathrm{Pd}}^{(1)} y_{\mathrm{Pd}}^{(2)}\, {}^0G_m^{\mathrm{Pd_3Pd}} + 3RT\left(y_{\mathrm{Al}}^{(1)} \ln y_{\mathrm{Al}}^{(1)} + y_{\mathrm{Pd}}^{(1)} \ln y_{\mathrm{Pd}}^{(1)}\right) + RT\left(y_{\mathrm{Al}}^{(2)} \ln y_{\mathrm{Al}}^{(2)} + y_{\mathrm{Pd}}^{(2)} \ln y_{\mathrm{Pd}}^{(2)}\right) + {}^{ex}G_m^{(\mathrm{Al_3Pd})}, \quad (1)$$

where $H_{(\mathrm{Al_3Pd})}^{\mathrm{SER}}$ is equal to $\left(3y_{\mathrm{Al}}^{(1)} + 1y_{\mathrm{Al}}^{(2)}\right) H_{\mathrm{Al}}^{\mathrm{SER}} + \left(3y_{\mathrm{Pd}}^{(1)} + 1y_{\mathrm{Pd}}^{(2)}\right) H_{\mathrm{Pd}}^{\mathrm{SER}}$. Terms $y_i^{(1)}$ and $y_i^{(2)}$ represent site fractions of the component i (i = Al, Pd) in the first and the second sublattices, respectively. Terms ${}^0G_m^{\mathrm{Al_3Al}}$, ${}^0G_m^{\mathrm{Al_3Pd}}$, ${}^0G_m^{\mathrm{Pd_3Al}}$, and ${}^0G_m^{\mathrm{Pd_3Pd}}$ correspond to Gibbs energies of hypothetic stoichiometric compounds Al_3Al, Al_3Pd, Pd_3Al, and Pd_3Pd which are formed when each sublattice is occupied by one component only (y_{Al}=1 or y_{Pd}=1). They are modeled using formulas of Gibbs energy for stoichiometric compounds. The term ${}^{ex}G_m^{(\mathrm{Al_3Pd})}$ is the excess Gibbs energy that can be expressed by formula:

$${}^{ex}G_m^{(\mathrm{Al_3Pd})} = y_{\mathrm{Al}}^{(1)} y_{\mathrm{Pd}}^{(1)} \left(y_{\mathrm{Al}}^{(2)} L_{\mathrm{Al,Pd:Al}}^{(\mathrm{Al_3Pd})} + y_{\mathrm{Pd}}^{(2)} L_{\mathrm{Al,Pd:Pd}}^{(\mathrm{Al_3Pd})}\right) + y_{\mathrm{Al}}^{(2)} y_{\mathrm{Pd}}^{(2)} \left(y_{\mathrm{Al}}^{(1)} L_{\mathrm{Al:Al,Pd}}^{(\mathrm{Al_3Pd})} + y_{\mathrm{Pd}}^{(1)} L_{\mathrm{Pd:Al,Pd}}^{(\mathrm{Al_3Pd})}\right). \quad (2)$$

The symbol L in Equation (2) stands for interaction parameters, writable as:

$$L_{\mathrm{Al,Pd}:k}^{(\mathrm{Al_3Pd})} = \textstyle\sum_{j=0}\left(a_j + b_j T\right)\left(y_{\mathrm{Al}}^{(1)} - y_{\mathrm{Pd}}^{(1)}\right)^j, \quad (3)$$

$$L_{k:\mathrm{Al,Pd}}^{(\mathrm{Al_3Pd})} = \textstyle\sum_{j=0}\left(a_j + b_j T\right)\left(y_{\mathrm{Al}}^{(2)} - y_{\mathrm{Pd}}^{(2)}\right)^j, \quad (4)$$

where $L_{\mathrm{Al,Pd}:k}$ and $L_{k:\mathrm{Al,Pd}}$ correspond to the interaction parameters between Al and Pd in one sublattice, whereas the other sublattice is occupied by the element k only (k = Al, Pd). To reduce the number of optimizing variables, it can be assumed:

$$L_{\mathrm{Al,Pd:Al}}^{(\mathrm{Al_3Pd})} = L_{\mathrm{Al,Pd:Pd}}^{(\mathrm{Al_3Pd})}, \quad (5)$$

$$L_{\mathrm{Al:Al,Pd}}^{(\mathrm{Al_3Pd})} = L_{\mathrm{Pd:Al,Pd}}^{(\mathrm{Al_3Pd})}. \quad (6)$$

3. Results

Thermodynamic descriptions of phases $\varepsilon_n/(Al_3Pd)$, λ-Al_4Pd, and γ-$Al_{21}Pd_8$ reflecting newly evaluated parameters are given in Table 1. In the calculations, thermodynamic parameters of remaining phases were taken from [7]. The phase diagram calculated subsequently by means of Thermo-Calc using modified parameters (Table 1) is shown in Figure 1a.

Table 1. Thermodynamic parameters of phases ε_n/(Al_3Pd), λ-Al_4Pd, and γ-$Al_{21}Pd_8$ evaluated in this work; energy terms are given in J/(mole of formula unit).

Phase	Thermodynamic model	Parameters
λ-Al_4Pd	$(Al)_4$:$(Pd)_1$	${}^0G_m^{Al_4Pd} = 4\,{}^0G_{Al}^{fcc} + {}^0G_{Pd}^{fcc} - 170737.44 + 51.21T$
γ-$Al_{21}Pd_8$	$(Al)_{21}$:$(Pd)_8$	${}^0G_m^{Al_{21}Pd_8} = 21\,{}^0G_{Al}^{fcc} + 8\,{}^0G_{Pd}^{fcc} - 1246650.94 + 275.07T$
ε_n/(Al_3Pd)	(Al%,Pd)$_3$(Al,Pd%)$_1$	${}^0G_m^{Al_3Al} = {}^0G_{Al:Al}^{(Al_3Pd)} = 4\,{}^0G_{Al}^{fcc} + 100000$
		${}^0G_m^{Pd_3Pd} = {}^0G_{Pd:Pd}^{(Al_3Pd)} = 4\,{}^0G_{Pd}^{fcc} + 41000$
		${}^0G_m^{Al_3Pd} = {}^0G_{Al:Pd}^{(Al_3Pd)} = 3\,{}^0G_{Al}^{fcc} + {}^0G_{Pd}^{fcc} - 156506.48 + 35.41T$
		${}^0G_m^{Pd_3Al} = {}^0G_{Pd:Al}^{(Al_3Pd)} = {}^0G_{Al}^{fcc} + 3\,{}^0G_{Pd}^{fcc} - 55000 + 20.0T$
		$L_{Al,Pd:Al}^{(Al_3Pd)} = -486700 + 28.7T$
		$L_{Al,Pd:Pd}^{(Al_3Pd)} = -486700 + 28.7T$
		$L_{Al:Al,Pd}^{(Al_3Pd)} = -25000 + 140.0T$
		$L_{Pd:Al,Pd}^{(Al_3Pd)} = -25000 + 140.0T$

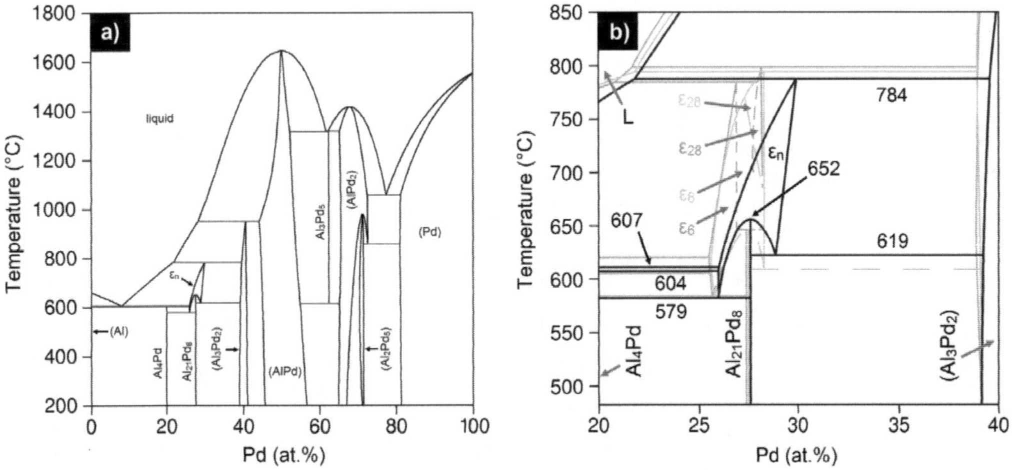

Figure 1. Calculated Al-Pd phase diagram proposed in this work (a), detail view in vicinity of ε_n/(Al_3Pd) area (b). In view (b), dark wider lines represent diagram proposed in this work, finer lines distinguished in colour represent experimental diagrams published in [3] (light-orange) and [5] (light-green). Characteristic concentrations to invariant reactions are given in text. For interpretation of figure in colour, reader is referred to web-version of article.

4. Discussion

The currently available versions of the Al-Pd phase diagram in vicinity of the ε_n/(Al_3Pd) area (i.e. published in [3], [5], and proposed in this work) are graphically compared in Figure 1b. In compliance with the latest published experiments related to the ε_6+ε_{28} area in the Al-Pd system [5], the single phase ε_n/(Al_3Pd) was thermodynamically described using the (Al%,Pd)$_3$(Al,Pd%)$_1$ two-sublattice thermodynamic model. In the calculated phase diagram shown in Figure 1, this phase occurs between 579°C and 784°C. The lower temperature limits of the ε_n/(Al_3Pd) occurrence are defined by eutectoid points at 26.0 at.% Pd (579°C) and 28.9 at.% Pd (619°C), whereas the upper temperature limit is defined by the peritectic point at 29.9 at.% Pd (784°C). According to the earlier published results [3], the ε_{28} phase is formed by peritectic reaction, L+δ/(Al_3Pd_2)$\leftrightarrow\varepsilon_{28}$, at 792°C and about 28.1 at.% Pd. In the present thermodynamic description, the peritectic point of the ε_n/(Al_3Pd) formation is at about 29.9 at.% Pd (Figure 1). The lowest temperature limit of the ε_n/(Al_3Pd) occurrence, 579°C, coincides with those previously experimentally determined [3] and calculated [7]. Between the calculated (784°C) and the earlier published (789°C [5]) upper temperature limits, there is the difference of 5°C only. The γ-$Al_{21}Pd_8$ phase was estimated to be stable up to 652°C. In the present work, the eutectoid reaction, ε_n/(Al_3Pd)$\leftrightarrow\gamma$-$Al_{21}Pd_8$+δ/(Al_3Pd_2), was calculated at 619°C (Figure 1), whereas it was not predicted in the phase diagram calculated previously [7]. The λ-Al_4Pd phase was calculated to be stable up to 604°C. However, thermodynamic parameters of it had to be slightly adjusted to reach the same formation temperature after involving the new thermodynamic model of ε_n/(Al_3Pd) in the calculations. As a consequence, the temperature of eutectic reaction, L$\leftrightarrow$(Al)+ε_n/(Al_3Pd), dropped to 607°C, Figure 1.

In the proposed calculated phase diagram, the peritectic point at 784°C, the eutectoid point at 619°C, and the eutectoid point at 579°C are all shifted to higher Pd-concentrations in comparison to the experimental diagrams published in [3] and [5] (see Figure 1b). As a consequence the whole ε_n/(Al_3Pd)-area is shifted to a bit higher Pd-concentrations as expected. Nevertheless, reasonable agreement was reached between the results of thermodynamic calculations and published experimental results.

5. Conclusions

The results obtained in a limited re-investigation of the Al-Pd system using CALPHAD method can be summarised as follows:

- ε_n/(Al_3Pd) was firstly described as a non-stoichiometric phase reflecting the two-sublattice thermodynamic model $(Al\%,Pd)_3(Al,Pd\%)_1$,
- thermodynamic parameters of the (Al_3Pd) phase were evaluated,
- thermodynamic parameters of phases λ-Al_4Pd and γ-$Al_{21}Pd_8$ were slightly adjusted,
- using the Thermo-Calc software, the Al-Pd phase diagram was re-calculated with respect to the newly involved (Al_3Pd).

Acknowledgements

The authors wish to thank the European Regional Development Fund (ERDF) for financial support of the project ITMS:26220120014 "Center for Development and Application of Advanced Diagnostic Methods in Processing of Metallic and Non-metallic Materials" funded within the Research & Development Operational Programme, the Grant Agency VEGA for the financial support under contracts 1/0018/15 and 1/0811/14, the Slovak Research and Development Agency for the financial support under the contracts APVV-0076-11 and APVV-15-0049, and the Slovak University of Technology for the financial contribution from the grant scheme number 1352 for supporting young researchers.

References

[1] McAlister A J 1986 *Bull. Alloy Phase Diagrams* **7** 368–74
[2] Massalski T B, Subramanian P R, Okamoto H and Kacprzak L 1990 (Eds.) *Binary Alloy Phase Diagrams*, Ohio, ASM International
[3] Yurechko M, Fattah A, Velikanova T and Grushko B 2001 *J. Alloys Comp.* **329** 173–81
[4] Okamoto H 2003 *J. Phase Equilib.* **24** 196
[5] Grushko B 2013 *J. Alloys Comp.* **557** 102–11
[6] Černičková I, Švec P, Watanabe S, Čaplovič Ľ, Mihalkovič M, Kolesár V, Priputen P, Bednarčík J, Janičkovič D and Janovec J 2014 *J. Alloys Comp.* **609** 73–9
[7] Li M, Li Ch, Wang F and Zhang W 2006 *Intermetallics* **14** 39–46

IOP Conf. Series: Journal of Physics: Conf. Series **809** (2017) 012009 doi:10.1088/1742-6596/809/1/012009

Experimental study of phases I and H in selected (56-60)Zn-(29-35)Mg-(8-13)Y alloys after long-term annealing at 400°C

Marek Adamech[1*], Ivona Černičková[1], Ľubomír Čaplovič[1], Pavol Priputen[1], Dušan Janičkovič[2], Jozef Janovec[3]

[1] Slovak University of Technology in Bratislava, Faculty of Materials Science and Technology in Trnava, Paulínska 16, 917 24 Trnava, Slovak Republic
[2] Slovak Academy of Sciences, Institute of Physics, Dúbravská 9, 845 11 Bratislava, Slovak Republic
[3] Slovak University of Technology in Bratislava, Vazovova 5, 812 43 Bratislava, Slovak Republic

*email: marek.adamech@stuba.sk

Abstract. Five (56-60)Zn-(29-35)Mg-(8-13)Y alloys (metal compositions are given in at.%) were investigated after annealing at 400°C for 150h. In the investigation, the scanning electron microscopy equipped with energy dispersive X-ray spectroscopy and the X-ray diffraction were used. The alloys were selected with the intention to collect experimental data about the (I+H) phase equilibrium at 400°C and to show trends applicable in a refinement of the related isothermal section of the Zn-Mg-Y phase diagram. As a result, a specific isothermal section at 400°C of Zn-Mg-Y phase diagram is presented, showing present experimental results and related thermodynamic and experimental data available in literature.

1. Introduction

Even though Zn-Mg-Y alloys have been studied sporadically since nineties [1-8], the findings do not facilitate any exhaustive description of this system. For instance, Tsai et al. [1] described a stable icosahedral quasicrystalline phase (I-phase) in the $Zn_{60}Mg_{30}Y_{10}$ alloy, earlier reported by Luo et al. [3]. Langsdorf et al. [6] showed that this phase solidifies through a peritectic reaction and coexists with brittle intermetallic compounds in the Zn-rich Zn-Mg-Y alloys with the Y contents higher than 4 at.%. Besides the I-phase (symmetry $Fm\bar{3}\bar{5}$ [1,7]), other binary or ternary phases can also appear in the Zn-Mg-Y system. For instance, the hexagonal H-phase $MgZn_3Y$ ($P6_3/mmc$ [9,10]), the cubic W-phase $Mg_3Zn_2Y_3$ ($Fm\bar{3}m$), the hexagonal Z-phase $Mg_{12}ZnY$ ($P6_3/mmc$ [1,11]), or the orthorhombic Mg_7Zn_3 (Immm [7,12]). Based on experimental results, Tsai et al. [1] proposed partial isothermal sections of the Zn–Mg–Y phase diagram at temperatures 427°C, 500°C and 600°C, containing homogeneity ranges of ternary phases Z, W, H and I. The calculated isothermal sections of this diagram at 600 °C, 500 °C and 400 °C were reported by Grobner et al. [7].

A lack of experimental and calculated data related to the (I+H) area evoked the current study. Five Zn-Mg-Y alloys compositionally close to the (I+H) area were investigated experimentally at 400 °C to collect data applicable in a refinement of the related isothermal section.

2. Experimental procedure

The investigated alloys, $Zn_{56.7}Mg_{34.3}Y_9$, $Zn_{57.9}Mg_{29.8}Y_{12.3}$, $Zn_{57.2}Mg_{29.9}Y_{12.9}$, $Zn_{56.8}Mg_{32.2}Y_{11}$, and $Zn_{59.6}Mg_{31.6}Y_{8.8}$, were prepared of pure metals in induction furnace under argon atmosphere. After casting, the samples were annealed at 400°C for 150 h and rapidly cooled in water to preserve their high-temperature microstructures. Particular microstructure constituents were observed and their metal compositions were determined by a JEOL JSM-7600F scanning electron microscope equipped with an Oxford Instruments X-max50 EDX spectrometer. An X-ray Panalytical Empyrean PIXCel 3D diffractometer with Bragg–Brentano geometry was used for phase identification. The characteristic CoKα1,2 radiation was generated at 40 kV and 40 mA. Measurements were done in the angular range 20° to 80°, with a step size of 0.0131° and the counting time 98 seconds per step.

3. Results

The microstructure constituents observed in the investigated alloys are listed in Table 1 by metal compositions and volume fractions. All of them were identified to be single-phase constituents comprising either I- or H-phases. In Figure 1 the microstructure of the $Zn_{57.9}Mg_{29.8}Y_{12.3}$ alloy is recorded, being characteristic for all the alloys. Similarly, the characteristic XRD pattern of double-phase microstructures (see Table 1) correspond to the $Zn_{57.9}Mg_{29.8}Y_{12.3}$ alloy (Figure 2). Besides the single-phase $Zn_{56.7}Mg_{34.3}Y_9$ alloy, the I-phase was found to dominate over the H-phase in all the alloys (Table 1).

Table 1. Metal compositions and other measured values of investigated alloys

alloy	colour of microstructure constituent	phase	atomic content in %			volume fraction in %
			Zn	Mg	Y	
$Zn_{56.7}Mg_{34.3}Y_9$	dark-grey	I	56.7 ± 0.1	34.3 ± 0.2	9.0 ± 0.2	100
$Zn_{57.9}Mg_{29.8}Y_{12.3}$	dark-grey	I	54.7 ± 0.4	35.3 ± 0.2	10.0 ± 0.2	80
	light-grey	H	65.3 ± 0.2	16.3 ± 0.2	18.4 ± 0.3	20
$Zn_{57.2}Mg_{29.9}Y_{12.9}$	dark-grey	I	53.8 ± 0.3	35.9 ± 0.2	10.3 ± 0.1	65
	light-grey	H	64.3 ± 0.2	17.4 ± 0.1	18.3 ± 0.3	35
$Zn_{56.8}Mg_{32.2}Y_{11}$	dark-grey	I	55.6 ± 0.4	34.7 ± 0.2	9.7 ± 0.2	95
	light-grey	H	62.5 ± 0.2	19.7 ± 0.4	17.8 ± 0.3	15
$Zn_{59.6}Mg_{31.6}Y_{8.8}$	dark-grey	I	59.5 ± 0.1	31.8 ± 0.2	8.7 ± 0.1	99
	light-grey	H	65.3 ± 0.4	16.4 ± 0.4	18.3± 0.3	1

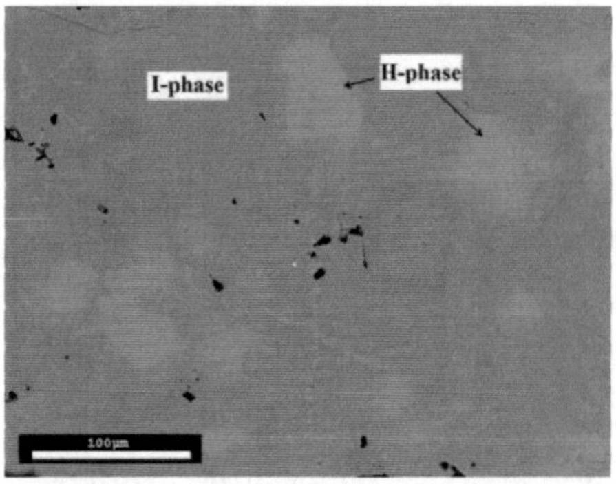

Figure 1. BEI/SEM image of microstructure of $Zn_{57.9}Mg_{29.8}Y_{12.3}$ alloy after annealing at 400°C for 150 h. Dark- and light-grey areas correspond to I- and H-phases, respectively. Black areas are pores.

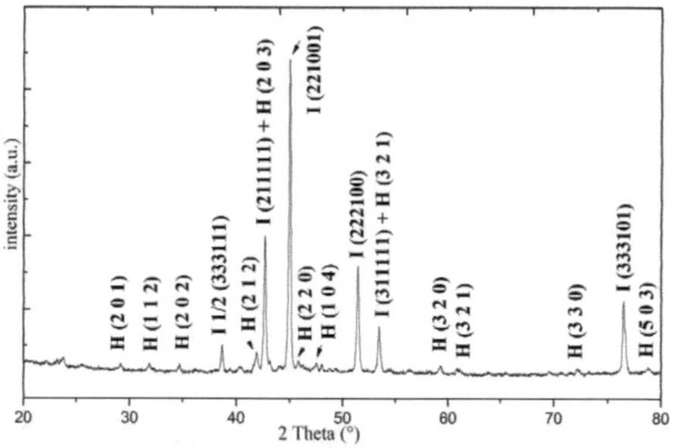

Figure 2. XRD pattern of $Zn_{57.9}Mg_{29.8}Y_{12.3}$ alloy after annealing at 400°C for 150 h. Corresponding phase and diffracting plane are assigned to each peak.

4. Discussion

The experimental data obtained at 400°C for phases I and H in this work were compared with those published in literature for compositionally comparable alloys annealed at the same temperature [7,8]. All the relevant data are illustrated in Figure 3 in the form of a specific isothermal section at 400°C of the Zn-Mg-Y phase diagram. The most complex experimental data for the Zn-Mg-Y system were presented by Tsai et al. [1]. Since part of them was obtained at 427°C, that is temperature slightly higher than 400°C, a mathematical extrapolation was used to estimate presumed shapes of I and H areas at 400°C. In this procedure, data from the isothermal sections at 500 °C and 427°C [1] were taken into account. For convenience, the linear regression was used based on the following function [13]:

$$y(x^*) = y_{k+1} + \frac{x^* - x_{k+1}}{x_k - x_{k+1}} (y_k - y_{k+1}), \quad (1)$$

where x_k and x_{k+1} represent independent variables of x (temperature in this work), and y_k and y_{k+1} stand for dependent variables of y (concentration). The symbol x* corresponds to the temperature to extrapolate. The obtained extrapolated areas for both the phases [1] were inserted into the specific isothermal section at 400°C, together with the thermodynamically predicted data [7], and present and earlier published experimental data [8], Figure 3.

An approximate homogeneity range of the H-phase was determined as (62-66)Zn-(16-20)Mg-(17-19)Y according to the data summarised in Table 1. The present experimental points are arranged in line, but they are localised at a higher Y level (about 18 at.%, Table 1) and in evidently shorter ranges of Zn and Mg contents (both about 5 at.%), as it was reported by Grobner et al. [7]. Similarly, the extrapolated homogeneity range of the H-phase ([1], Figure 3) enables variabilities in Zn and Mg contents inside ranges both of about 6 at.%. If compared to the present experimental data with the extrapolated data [1], the H-area proposed in the present work is shifted to higher Mg and Y contents, and to lower Zn contents. The shifts are not dramatic and could be caused by annealing conditions. For instance, the annealing time plays in this matter an important role, because shorter times lead to an insufficient homogenization and longer times can evoke evaporation of light elements from annealed samples. In the present work, the optimal duration of annealing was determined experimentally as 150 h. Annealing times used in the work [1] were not specified.

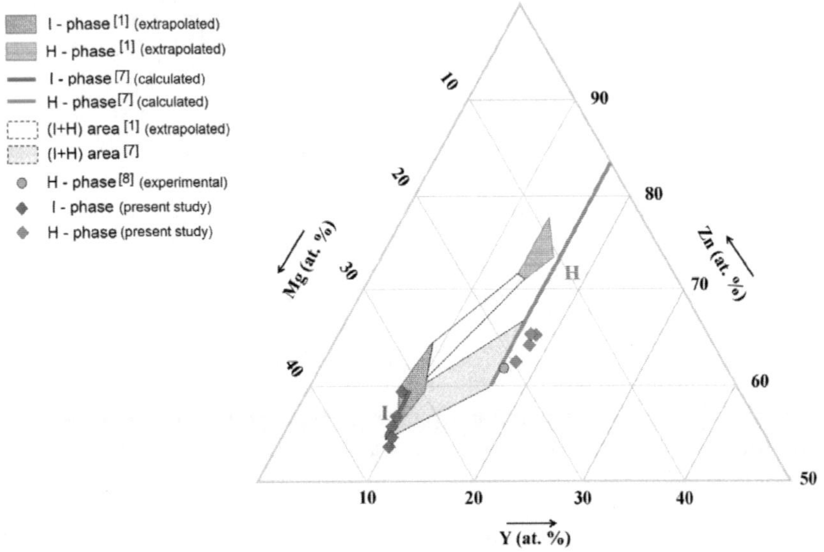

Figure 3. Specific isothermal section at 400°C of Zn-Mg-Y phase diagram to compare present experimental results with those available in literature (see legend for more detail description).

According to the present experimental results, an approximate homogeneity range of the I-phase was determined as (53-60)Zn-(31-36)Mg-(8-11)Y (Table 1). Positions of the I-area specified by extrapolated [1], predicted [7], and present experimental data correlate with each other very well (Figure 3). The former position is shifted only slightly to higher Zn and lower Mg contents. The results obtained in this work confirmed the earlier published finding about the compositionally limited I-area [1,7].

The thermodynamically predicted and the extrapolated (I+H) areas have no common point in the specified isothermal section (Figure 3). The experimental results both obtained in the present work and presented by Hamaya et al. [8] correlate better with the (I+H) position predicted thermodynamically by Grobner et al. [7].

5. Conclusions

The results obtained in the investigation of alloys $Zn_{56.7}Mg_{34.3}Y_9$, $Zn_{57.9}Mg_{29.8}Y_{12.3}$, $Zn_{57.2}Mg_{29.9}Y_{12.9}$, $Zn_{56.8}Mg_{32.2}Y_{11}$, $Zn_{59.6}Mg_{31.6}Y_{8.8}$ after annealing at 400°C for 150 h can be summarised as follows:

- The I-phase identified in all the investigated alloys was found to fall within the (53-60)Zn-(31-36)Mg-(8-11)Y area. This finding correlates with the data available in literature very well and confirms the suggestion about the compositionally limited I-area.
- The H-phase was found in double-phase microstructures only, coexisting with the I-phase. Its metal composition ranges within the (62-66)Zn-(16-20)Mg-(17-19)Y area.
- A specific isothermal section at 400°C of Zn-Mg-Y phase diagram is presented, showing both the present experimental results and the related results available in literature. It follows from the section that the (I+H) area determined experimentally in this work correlates with that predicted thermodynamically by Grobner et al. very well.

Acknowledgemets

The authors wish to thank the European Regional Development Fund (ERDF) for financial support of the project ITMS:26220120014 "Center for Development and Application of Advanced Diagnostic Methods in Processing of Metallic and Non-metallic Materials" funded within the Research & Development Operational Programme, the Grant Agency VEGA for the financial support under contract 1/0018/15, the Slovak Research and Development Agency for the financial support under contracts APVV-0076-11 and APVV-15-0049, and Grant 1352 for Support of Young Researchers.

References

[1] Tsai A P, Murakami Y, Niikura A 2000 *Philos. Mag A.* **80** 1043-1054
[2] Padezhnova E M, Melnik E V, Miilzevskiz R A, Dobatkina T V, Kinzhibalo V V 1982 *Russ. Metall.* **4** 185-188
[3] Luo Z, Zhang S, Tang Y, Zhao D 1993 *Scr. Metall. et Mater.* **28** 1513–1518
[4] Lee J Y, Kim D H, Lim H K, Kim D H 2005 *Mater. Lett.* **59** 3801–3805
[5] Shao G, Varsani V, Fan Z 2006 *CALPHAD* **30** 286–295
[6] Langsdorf A, Ritter F, Assmus W 1997 *Phil. Mag. Lett.* **75** 381–388
[7] Grobner J, Kozlov A, Fang X Y, Geng J, Nie J F, Schmid R 2012 *Acta Mater.* **60** 5948–5962
[8] Hamaya A, Horiuchi T, Oohira Y, Minamoto S, Miura S, Sakaguchi N 2013 *Mater. Trans.* **54** 641
[9] Singh A, Watanabe M, Kato A, Tsai A P 2004 *Scripta Mater.* **51** 955–960
[10] Deng D W, Kuo K H, Luo Z P, Miller D J, Kramer M J, Dennis K W 2004 *J. Alloys Compd.* **373** 156–160
[11] Li M R, Kuo K H 2007 *J. Alloys Compd.* **432** 81–89
[12] Higashi I, Shiotani N, Uda M 1981 *J. Solid State Chem.* **36** 225–233
[13] Berry J, Nocliffe A, Humble S 1989 *Introductory mathematics through science applications*, Cambridge University Press, New York, pp. 560

A new superlattice structure in the $Al_{2.75}Ir$ and $Al_{2.63}Rh$ 1/0 approximants

Kazue Nishimoto[1], Natsumi Yamaguchi[2], Ryuji Tamura[2]

[1]Institute of Multidisciplinary Research for Advanced Materials (IMRAM), Tohoku University, 2-1-1, Katahira, Sendai-shi, Miyagi, 980-8577, Japan

[2]Department of Materials Science and Technology, Tokyo University of Science, 6-3-1 Niijuku, Katsushika-ku, Tokyo, 125-8585, Japan

nishi-mt@tagen.tohoku.ac.jp

Abstract. We investigate the superstructure of binary $Al_{2.75}Ir$ and $Al_{2.63}Rh$ 1/0 approximants by X-ray diffraction method and transmission electron microscopy. $Al_{2.75}Ir$ possesses a $2a \times 2a \times 2a$ *C*-centred monoclinic lattice with space group *C2* or *Cm* or *C2/m*. $Al_{2.63}Rh$ possesses a $2a \times 2a \times 2a$ face centred lattice with space group *F23* or $Fm\bar{3}$.

1. Introduction

Binary $Al_{2.75}Ir$ and $Al_{2.63}Rh$ are 1/0 approximants of ternary Al-TM quasicrystals. The structure of $Al_{2.75}Ir$ and $Al_{2.63}Rh$ have been investigated by XRD, and $Al_{2.75}Ir$ and $Al_{2.63}Rh$ possess a primitive cubic lattice with space group *P23* (the lattice parameter $a = 7.674(1)$ Å and $a = 7.6692(1)$ Å, respectively) [1]. Fig.1 shows schematic crystal structure of (a) $Al_{2.75}Ir$ and (b) $Al_{63.3}Cu_{12.3}Ir_{24.4}$ phases. In the case of $Al_{2.75}Ir$, the primitive cubic lattice is composed by Ir_{12} (green circle) icosahedron containing central Ir atom and 9~10 Al atoms (small yellow circle). It should be noted that the Al sites inside the Ir icosahedron are partially occupied with a pentagon-dodecahedral like distribution and surround central Ir atom. In contrast, the structure of ternary Al-Cu-TM 1/0 approximants are different from the binary Al-TM 1/0 approximants. The ternary Al-Cu-TM 1/0 approximants possess the double $Fm\bar{3}$ lattice with the lattice constant =15.3844(2) Å $(2a \times 2a \times 2a)$ due to the ordered Cu site inside of the TM icosahedron [2, 3]. Therefore, the structure of Al-Cu-TM 1/0 approximant is a $2a \times 2a \times 2a$ FCC (space group $Fm\bar{3}$).

In the case of Al-Ir, the order-disorder phase transition of Al inside TM icosahedron is predicted by theoretical calculation using pair potentials fitted to an ab initio database [4]. Moreover, Mihalkovič reported that the ordered phase of $Al_{2.75}Ir$ is possibly insulate phase. Similarly, for $Al_{2.73}Ir$, very weak superlattice reflections were reported in the selected area electron diffraction patterns (SAED) [5]. Hence, it is necessary to investigate the superlattice structure of binary Al-TM(TM = Ir, Rh) 1/0 approximants.

In the present work, we have investigated the superstructure and the composition dependence of the $Al_{2.75}Ir$ and $Al_{2.63}Rh$ by the powder X-ray diffraction (XRD) method and transmission electron microscopy (TEM) study to understand the behaviour of atoms inside TM icosahedral clusters.

2. Experimental

Al-TM(TM=Ir, Rh) with composition ranging from 70 to 75 % Al were melted by arc method under an argon atmosphere (Al (99.99%), Rh (99.9%) and Ir (99.9%)). And the samples were annealed at 1373K for about 72 hours or 723K for 3 days in the quarts tube under the argon atmosphere, followed by water quenching. Powder X-ray diffraction experiments were carried out using Cu Kα to examine the phase constitution. Additionally, TEM study were performed by JEOL -TKP2 or TOPCON-002b operating at 200kV. The samples for TEM were crushed by agate mortar and dispersed on a copper micro-grid mesh.

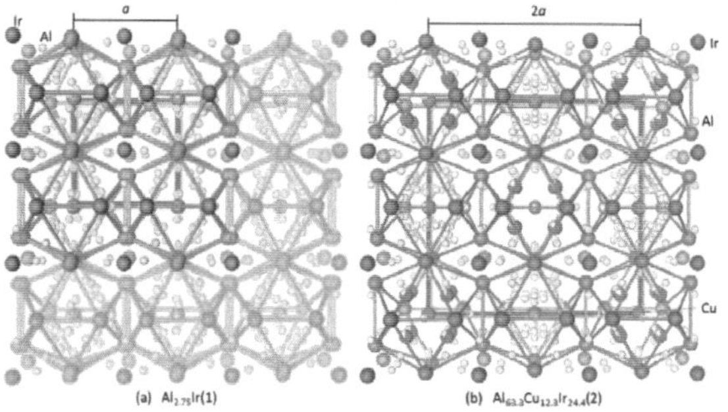

Fig. 1. Crystal structures of (a) $Al_{2.75}Ir$ with an $a \times a \times a$ lattice (space group *P23*) [1] and (b) $Al_{63.3}Cu_{12.3}Ir_{24.4}$ with an $2a \times 2a \times 2a$ lattice (space group $Fm\overline{3}$)[2]. The Al sites (small circle) inside the Ir icosahedron are partially occupied with a pentagon-dodecahedral-like distribution.

3. Results and discussions

3.1. XRD

Fig. 2 shows XRD patterns of (a) $Al_{73}Ir_{27}$, (b) $Al_{73.5}Rh_{26.5}$ and (c) $Al_{72.5}Rh_{27.5}$. The most peaks are indexed by an $a \times a \times a$ lattice with reported space group *P23*, but weak superlattice peaks appeared. In Fig. 2, superlattice peaks are indicated by black circles and open circles. For $Al_{73}Ir_{27}$, the superlattice peaks can be indexed as $2a \times 2a \times 2a$ lattice. In addition, the peaks parallel to the (100) seem to be split into two peaks, when the $K\alpha_2$ removed. Then, it is possible that the superlattice structure is monoclinic.

On the other hand, for Al-Rh, $Al_{2.63}Rh$ phase is observed in the composition range of Al70~74.5%. Indexed as a $2a \times 2a \times 2a$ lattice, the indexes ($h_F k_F l_F$) satisfied the reflection condition ($h_F k_F l_F$; h_F+k_F, k_F+l_F, l_F+h_F = even. It is possible that the superlattice structure possesses FCC lattice with a lattice constant a_F (= $2a$) = 15.340Å. Additionally, superlattice peaks disappeared in the poorer Al compositions below Al72.5%. The composition dependence on the superstructure suggests that the number of Al inside the TM icosahedron contribute to the superstructure.

3.2. TEM

3.2.1 $Al_{2.63}Rh$

Figure 3 shows selected area diffraction patterns of $Al_{72.5}Rh_{27.5}$ taken along [100], [110] and [111]. We observed the superlattice reflections at h/2 k/2 l/2 positions indicated by open triangle in the [110] SAED pattern. In the [111], there are no superlattice reflections, suggesting that the superlattice structure is cubic lattice. All the superlattice reflections can be indexed by considering a $2a \times 2a \times 2a$ lattice ($h_F k_F l_F$), and satisfy the FCC reflection condition ($h_F k_F l_F$; h_F+k_F, k_F+l_F, l_F+h_F = 2n). Therefore, the superlattice of $Al_{2.63}Rh$ can be explained as $2a \times 2a \times 2a$ *F23* or $Fm\overline{3}$.

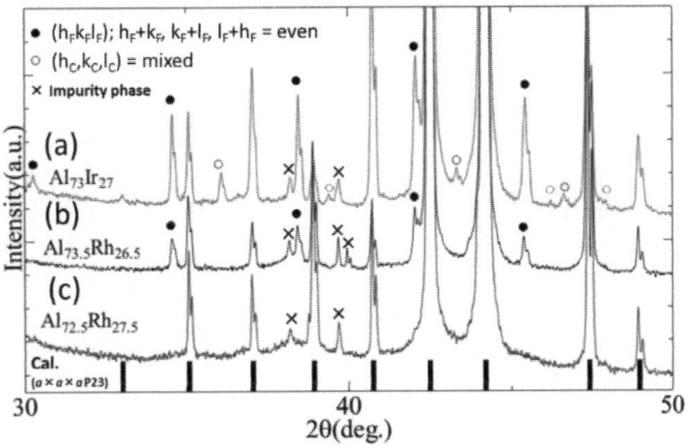

Fig. 2. XRD patterns of (a) $Al_{73}Ir_{27}$, (b) $Al_{73.5}Rh_{26.5}$ and (c) $Al_{72.5}Rh_{27.5}$.

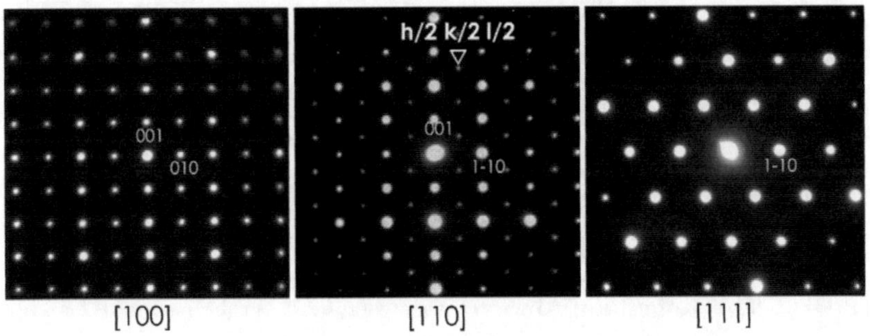

Fig 3. Selected area electron diffraction patterns of Al-Rh taken along [100], [110], [111] and (b) a bright field TEM image.

3.2.2 $Al_{2.75}Ir$

Figure 4 presents SAED patterns of $Al_{73}Ir_{27}$ taken along [100], [001]. [110]. [011] and [111]. We observed the superlattice reflections at h/2 k/2 l/2, h/2 k/2 0 and 0 0 l/2 positions indicated by open and filled triangles. Thus, all the superlattice reflections can be indexed by considering a $2a\times2a\times2a$ lattice ($h_Ck_Cl_C$). In the case of [111], the superlattice reflections appeared at the point to break the cubic 3-fold symmetry. It indicates that the superlattice structure are not cubic lattice. Considering the XRD result, it is possible that the $Al_{2.75}Ir$ possesses a monoclinic lattice. Moreover, all reflections satisfy the *C*-centred lattice reflection condition ($h_Ck_Cl_C$; $h_C+k_C=2n$). Therefore, the superlattice structure of $Al_{2.75}Ir$ is $2a\times2a\times2a$ *C2* or *Cm* or *C2/m*.

Fig 5 shows TEM images of $Al_{7.25}Ir$ annealed at 723K for 3 days ((a, b) and (c) a HRTEM image. (b) and (c) were taken along [100]). In Fig. 5(a), dark contrast line indicates domain boundary of ordered phase due to the structural phase transition. The structural phase transition might be same mechanism of order-disorder phase transition of Cd_6M 1/1 approximants at about 100 K [6]. The domain structure is the rectangular shape of about a few 100 nm. Also, in the SAED taken from domains (Fig.5b), there are diffuse streaks along [001]. Additionally, many dark parallel lines generated by stacking faults can be recognized thickly at intervals of a few dozen nm parallel to (001) in the domains (Fig. 5b). It implies that the stacking of ordered clusters tend to easily disarranged to [001]. To obtain the insulate phase for Al-Ir phase, it is interesting to investigate the electronic state of superlattice structures and lattice defects.

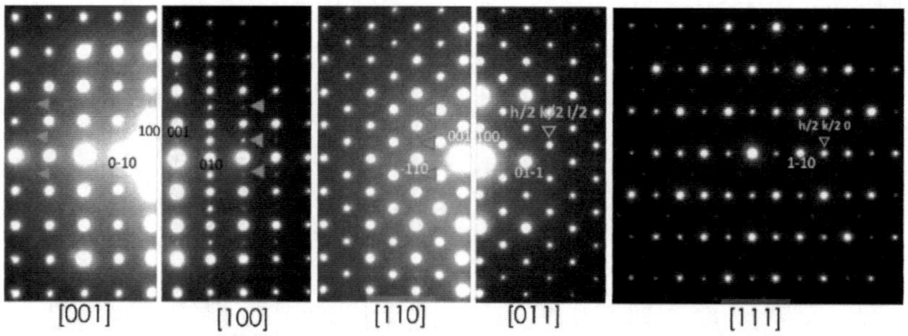

Fig 4. Selected area electron diffraction patterns of $Al_{7.25}Ir$ taken along [100], [110], [011] and [111] annealed at 1373K for 72h.

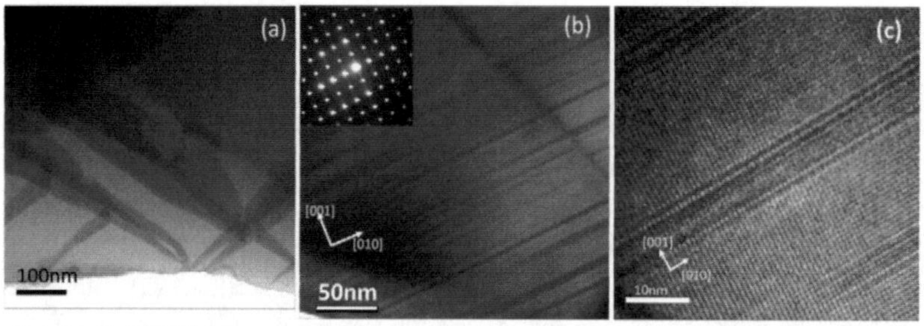

Fig. 5 (a) Bright field TEM image of Al2.75Ir annealed at 723k for 3 days. (b) BF and (c) lattice images were taken along [100].

In this work, we reported the new superlattice structure of binary $Al_{2.75}Ir$ and $Al_{2.63}Rh$ phases, which was effected by number of Al atoms in side TM(TM = Ir, Rh) icosahedron cluster. $Al_{2.75}Ir$ possesses a $2a \times 2a \times 2a$ monoclinic lattice with space group *C2* or *Cm* or *C2/m*. $Al_{2.63}Rh$ possesses a $2a \times 2a \times 2a$ face centred lattice with space group *F23* or $Fm\overline{3}$. The composition dependence of superlattice reflections suggest that the superlattice strucuture is determined by the number of ordered Al atoms inside TM icosahedron.

Acknowledgements

TEM observations were performed by JEOL-TKP02 at Prof. Masami Terauchi's laboratory, IMRAM, Tohoku University.

References

[1] Grin Y, Peters K, Burkhardt U, Gotzmann K and Ellneret. M 1997 *Z. Kristallogr.* **212** 439
[2] Dshemuchadse J, Kuczera P and Steurer W 2013 *Intermetallics* **23** 337
[3] Grushko B, Gwóźdź J and Yurechko M 2000 *J. Alloys. Comp.* **305** 219
[4] Oishi A, Nishimoto K and Tamura R 2009 *Z. Kristallogr.* **224** 115
[5] Mihalkovič M and Henley C L 2013 *Phys. Rev. B* **88** 064201
[6] Nishimoto K, Sato T and Tamura R 2013 *J. Phys. Cond. Mat.* **25** 235403

Synthesis characterization and hydrogenation behaviour of as quenched $Ti_{41.5+x}Zr_{41.5-x}Ni_{17}$ (x=0, 3.5, 11.5 and 13.5) nano quasicrystalline ribbons

Rohit R Shahi[1], T.P. Yadav[2], M.A. Shaz[2], O.N. Srivastava[2]

[1]Department of Physics Motilal Nehru National Institute of Technology Allahabad-211004 India,
[2]Department of Physics, Banaras Hindu University Varanasi-221005, India
Email; rohitrshahi@gmail.com

Abstract. The present study describes the synthesis characterization and hydrogen storage behavior of $Ti_{41.5+x}Zr_{41.5-x}Ni_{17}$ (x=0, 3.5, 11.5 and 13.5) nano quasicrystalline ribbons synthesized through rapid solidification technique. The nano quasicrystalline ribbons of alloys have been synthesized at Cu-wheel speed of 45 m/sec. The investigation also describes the effect of different compositions of Ti and Zr on the structure, microstructure and their correlation with hydrogen storage characteristics of as quenched nano quasicrystalline ribbons. It has been found that the hydrogen storage capacity of these ribbons increases with increasing the content of Ti in i-phase ribbons.

1. Introduction

It is now well known that storage of hydrogen is main challenge in developing hydrogen economy. Hydrogen can be stored at moderate temperature and pressure in intermetallic compounds and formed metal hydrides [1, 2]. Intensive research has been conducted in last few decades on metal hydrides to improve the hydrogen storage capacity, kinetics, thermal properties and cycling behavior [3]. In this series quasicrystalline alloys have lately be confirmed to store large amount of hydrogen reversibly because of the availability of higher number of interstitial voids [4]. After the discovery of a stable icoshedral phase (i-phase) in Ti-Zr-Ni system, this system received attention as hydrogen storage materials [5, 6]. The hydrogen uptake capacity of i - phase is found to be higher as compared to other crystalline phases due the shell structure of i–phase which provide higher density of interstitial voids as compared to crystalline phases. For the Ti-Zr-Ni system the presence of Ti/Zr enhances the hydrogen uptake capacity due to the better affinity of Ti and Zr towards hydrogen. Moreover, the presence of Ni enhances the dissociation of hydrogen molecule to hydrogen atom at the surface and it often creates a pathway for H diffusion through a surface oxide layer.

Ti-based IQC alloys can be synthesized by mechanical alloying followed by annealing and also by rapid solidification technique. Several studies confirmed that the phase formation and properties of these alloys strictly affected by the processing parameter and variation of stoichiometric ratio [7,8]. We have also been investigated the effect of processing parameter by changing the quenching rate on the phase formation and hydrogen storage characteristics of $Ti_{45}Zr_{38}Ni_{17}$ IQC ribbons synthesized through rapid solidification technique and further investigated their hydrogenation characteristics [9]. It has been found that hydrogen uptake capacity as well as the absorption kinetics of the IQC ribbons increases with increasing the quenching rate at which the ribbons were synthesized due to the formation of nano grains of i-phase. In the present report, we have investigated the effect of composition on the phase formation, microstructure and their hydrogen storage behavior of $Ti_{41.5+x}Zr_{41.5-x}Ni_{17}$ (x= 0, 3.5, 11.5 and 13.5) i-phase ribbons. We have synthesized nano quasi-crystalline ribbons of $Ti_{41.5+x}Zr_{41.5-x}Ni_{17}$ at wheel speed of 45m/s and investigated their hydrogen storage characteristics.

2. Experimental Details

Alloy ingots of desired stoichiometry were prepared by RF induction melting of pure Ti (99.99%), Zr (99.99%) and Ni (99.99%) metal in graphite crucible under argon atmosphere. About 10g as cast ingots were melted 4 times to achieve chemical homogeneity. Further the alloy ingot was subjected to rapid solidify by using Cu-wheel of diameter 14 cm and at tangential speed of 45m/s . The formed ribbons have the thickness of 40 to 50 μm and length 30 to 40 cm. The synthesized ribbons were structurally characterized by X–ray diffraction (XRD) (Rigaku) equipped with CuKα radiation. The XRD data is further used to evaluate the lattice strain and crystallite size of the synthesized ribbon through pseudo Viot function analysis [10]. The microstructure and rotational symmetry of the synthesized ribbons were investigated through transmission electron microscopy (TEM) (FEI-Tecnai 20 G^2). The TEM sample were prepared by thinning of ribbons through an electro-polishing unit (Tenupol-5) using electrolyte 1:4 ratios of $HClO_4$ and methanol at 30°C.The surface morphology and composition of the synthesized ribbons were characterized by scanning electron microscopy (SEM) (Quanta 200) equipped with EDAX facility. The hydrogen sorption behavior was examined though the computerized pressure-concentration-temperature measurement system (AMC-USA) having the facility to measured the absorption desorption kinetics of the samples.

3. Results and Discussion

We have already investigated the effect of quenching rate on phase formation and hydrogen storage behavior of $Ti_{45}Zr_{38}Ni_{17}$ IQC ribbons and found that 45m/s wheel speed is sufficient for the synthesis of nano quasicrystalline phase [7]. Therefore, in the present study we have synthesized ribbons of $Ti_{41.5+x}Zr_{41.5-x}Ni_{17}$ (x=0, 3.5, 11.5 and 13.5) at wheel speed of 45m/s. Fig. 1 represents XRD patterns of different $Ti_{41.5+x}Zr_{41.5-x}Ni_{17}$ ribbons (for x=0, 3.5,11.5 and 13.5) at wheel speed of 45m/s. A hump like region appeared in all the XRD patterns which are coming from the XRD sample holder. The XRD pattern of $Ti_{41.5}Zr_{41.5}Ni_{17}$ (as shown in fig 1(a)) having peaks index with i-phase (indexed according Bancel scheme [11]). Further if we increase the content of Ti from 41.5 to 45 at% and correspondingly decrease the content of Zr from 41.5 to 38 at% the XRD pattern of $T_{45}Zr_{38}Ni_{17}$ ribbons reveals the presence of peaks which are indexed with i-phase similar to the case of $T_{41.5}Zr_{41.5}Ni_{17}$. However, the peak of i-phase becomes broadened as compared to first sample and shifted towards lower 2Θ values from 36.93 to 36.37 for $Ti_{45}Zr_{38}Ni_{17}$. As we further increase the content of Ti from 45 to 53 (x=11.5) the peaks correspond to i-phase become broadened and shifted towards higher 2Θ values (as shown in fig 1(c)). This may happen due to higher concentration of Ti and lower concentration of Zr. Further by increasing the content of Ti (from 53 to 55 at %) (x=13.5) and correspondingly decrease the content of Zr the position of most intense peak is same as that of the case for $Ti_{53}Zr_{30}Ni_{17}$ (as shown in fig. 1(d)). In order to understand the effect of different concentration of Ti and Zr on crystallite size and lattice strain, we have evaluated these by pseudo Viot function analysis of XRD profile. In this regard, the most intense peak of i-phase has been fitted by considering Gaussian and Lorenzian profile function. The Gaussian and Lorenzian widths were separated out and as we know these are correspond to lattice strain and crystallite size respectively. The evaluated lattice strain and crystallite size for $Ti_{41.5+x}Zr_{41.5-x}Ni_{17}$ i-phase ribbons are summarized in Table1 and found that as we increase the content of Ti, the crystallite size of the synthesized i-phase is decreased.

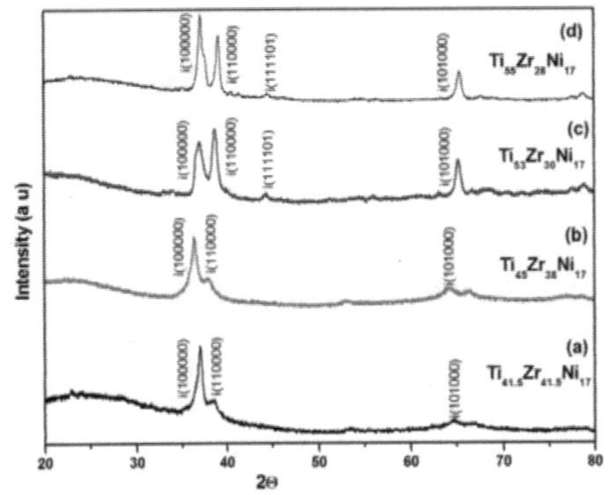

Table 1; for lattice strain and crystallite size

Sample	Lattice strain (±0.05)	Crystallite Size (± 0.5nm) (nm)
$Ti_{41.5}Zr_{41.5}Ni_{17}$	0.185	30.7
$Ti_{45}Zr_{38}Ni_{17}$	0.261	27.5
$Ti_{53}Zr_{30}Ni_{17}$	0.310	20.0
$Ti_{55}Zr_{28}Ni_{17}$	0.210	15.7

Fig.1 XRD patterns of (a) $Ti_{41.5}Zr_{41.5}Ni_{17}$, (b) $Ti_{45}Zr_{38}Ni_{17}$, (c) $Ti_{53}Zr_{30}Ni_{17}$ and (d) $Ti_{55}Zr_{28}Ni_{17}$ ribbon synthesized at wheel speed of 45 m/s

The TEM micrograph of $Ti_{41.5+x}Zr_{41.5-x}Ni_{17}$ (x=0,3.5, 11.5 and 13.5) i-phase ribbons synthesized at 45m/s are shown in Fig.2. Fig.2 (a) represents the TEM micrograph of $Ti_{41.5}Zr_{41.5}Ni_{17}$ ribbon reveals the formation grains of size 200 nm. The corresponding SAED patterns show the characteristics 5–fold and 2–fold symmetry of i-phase at different tilt (shown in fig. 2(a)). The synthesized ribbon of $Ti_{45}Zr_{38}Ni_{17}$ having grains ranges from 100–150 nm (as shown in Fig. 2b). The corresponding SAED pattern at different tilt shows the 5–fold and 2–fold symmetry of i-phase. Fig.2(c) represents the TEM micrograph of $Ti_{53}Zr_{30}Ni_{17}$ ribbon reveals the formation of i-phase grains of size ranges from 20-50 nm. The representative SAED pattern reveals diffuse ring along with the distorted 5-fold symmetry which is appearing due to the formation of nano grains. For the case of $Ti_{55}Zr_{28}N_{17}$ the microstructure remains the similar as $Ti_{53}Zr_{30}Ni_{17}$ ribbons. However, the grain size of i-phase is further decreased even by slight change of Ti content (53 to 55 at%). Hence it confirms that as we increase the content of Ti the grain size of the synthesized ribbon decreases. The corresponding SAED pattern shows more distorted 5-fold and 2-fold symmetry. The EDAX analysis were carried out over synthesized ribbons through SEM (Quanta–200) did not reveal the oxygen content (not shown here). The alloy composition has been found close to the nominal stoichiometry taken.

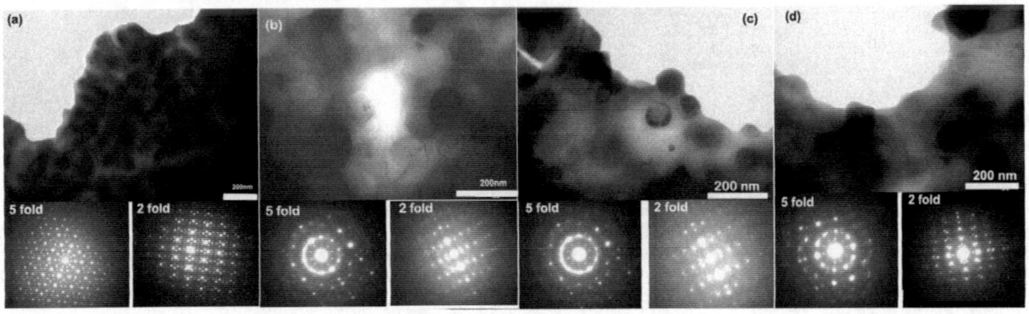

Fig.2 TEM micrograph and corresponding 5-fold, 2-fold symmetry of i-phase of (a) $Ti_{41.5}Zr_{41.5}Ni_{17}$, (b) $Ti_{45}Zr_{38}Ni_{17}$, (c) $Ti_{53}Zr_{30}Ni_{17}$ and (d) $Ti_{55}Zr_{28}Ni_{17}$ ribbons.

The ribbons kept in reactor were evacuated up to 10^{-3} torr through vacuum pump and then heated up to 320°C and charged with 7MPa of hydrogen pressure. In order to activate the sample, the reactor is evacuated for 3–4 times after charging with hydrogen. The hydrogen absorption kinetics was measured after the activation of synthesized ribbons. Fig. 3(a) shows the absorption characteristics of $Ti_{41.5+x}Zr_{41.5-x}Ni_{17}$ ribbons synthesized at wheel speed of 45m/s. The hydrogen uptake capacity is found to 2.15, 2.4, 2.6 and 2.7 wt% of hydrogen within 5.5hr for $Ti_{41.5}Zr_{41.5}Ni_{17}$, $Ti_{45}Zr_{38}Ni_{17}$, $Ti_{53}Zr_{30}Ni_{17}$ and $Ti_{55}Zr_{28}Ni_{17}$ i-phase ribbons respectively. The hydrogen desorption characteristics of hydrogenated ribbons were investigated through hydrogen desorption kinetics measurement performed at 350°C and 0.5 atm hydrogen pressure. Fig.3 (b) represents the representative desorption kinetics plot of nanocrystalline ribbons. From the Fig. it is clear that as we increase the Ti content desorption kinetics increases. Thus hydrogen absorption as well as desorption characteristics of synthesized nano quasicrystalline ribbons increases as we increase the content of Ti this may occur due to the formation of more finer grains which enhances the density of active grains for the interaction with hydrogen [9]. Moreover, the hydrogen uptake capacity also may increase due to the fact that the density of Ti (4.43 g/cm^3) is less as compared to Zr ($6.49 g/cm^3$). Thus as the content of Ti increases and corresponding decrease in the content of Zr, the overall density of the ribbons is going to reduce and hence the overall hydrogen uptake capacity of the synthesized ribbon is increased.

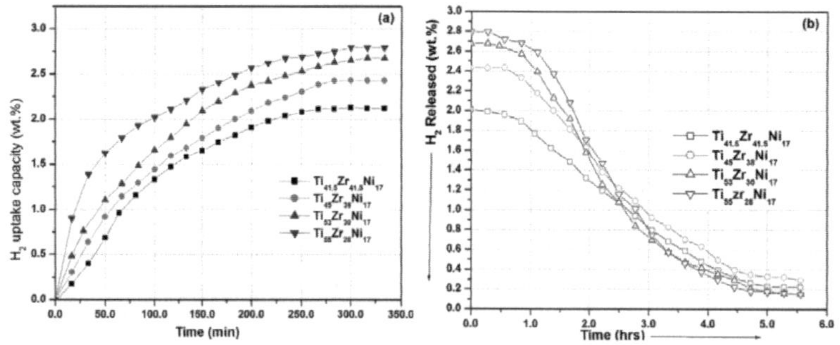

Fig.3 (a) Hydrogen absorption kinetics plot at 320 °C & 7MPa H_2 pressure and (b) hydrogen desorption measurement at 350 °C &0.5atm pressure for $Ti_{41.5+x}Zr_{41.5-x}Ni_{17}$ (x=0, 3.5, 11.5 and 13.5).

Conclusion

We have studied the variation of the composition on structure, microstructure and their correlation for hydrogen storage characteristics of $Ti_{41.5+x}Zr_{41.5-x}Ni_{17}$ i-phase ribbons. Nano quasi crystalline ribbons of i-phase have been synthesized through rapid solidification technique at wheel speed of 45m/s. it is found that as we increase the content of Ti and correspondingly decrease the Zr content in i-phase ribbons the formation of finer grains take place. Due to which the hydrogen uptake capacity and the kinetics of ab/desorption of hydrogen for the synthesized i-phase ribbons is increased.

Acknowledgment

One of the authors (Dr. R. R. Shahi) would like to acknowledge the financial assistance from DST-INSPIRE Project (IFA-12-PH-43) and XRD characterization facility of CIR-MNNIT Allahabad India.

References

[1] Schlapbach L and Züttel 2001 *Nature* **414** 353.
[2] Srivastava ON, Yadav TP, Shahi RR, Pandey SK, Shaz MA and Bhatnagar A 2015 Proc. *Indian Natn. Sci. Acad.* **81** 915.
[3] Sakintuna B and Hirscher M 2007 *Int. J. of Hyd .Energy* **32** 1121.
[4] Takasaki A and Kelton KF 2006 *Int. J. of Hyd .Energy* **31** 183.
[5] Kelton KF, Kim WJ and Stroudb RM 1997 *Appl. Phys. Lett.* **70** 24.
[6] Stroud RM, Viano AM, Gibbons PC and Kelton KF **1996** *Appl. Phys. Lett.* **69** 20.
[7] Lefaix H, Vermaut P, Janickovic D, Svec P, Portier R and Prima F **2009** *J Alloy Com.* **1-2** 168.
[8] Davis JP, Majzoub EH, J.M. Simmons and Kelton KF 2000 *Mat Sci and Eng* **294–296** 104.
[9] Shahi RR, Yadav TP, Shaz MA, Srivastava ON, Smaalen SV 2011, *Int. J hyd. Energy* **36** 592.
[10] de Keljser Th H, Langford JI, Mittemeijer EJ, Vogels ABP 1982 *J.Appl. Cryst.* 15, 308.
[11] Bancel PA, Heiney PA 1985 *Phys. Rev. Letters* 54, 2422.

Magnetic-field dependence of Schottky specific heat in the 1/1 Zn-Sc-Tm approximant

J Dolinšek

J. Stefan Institute and University of Ljubljana, Faculty of Mathematics and Physics, Jamova 39, SI-1000 Ljubljana, Slovenia

E-mail: jani.dolinsek@ijs.si

Abstract. The analysis of low-temperature specific heat of rare-earth (RE)-containing quasicrystals and periodic approximants and consequent interpretation of their electronic properties in the $T \to 0$ limit is frequently hampered by the Schottky effect, where crystalline electric fields lift degeneracy of the RE-ion Hund's rule ground state and introduce additional contribution to the specific heat. When the low-temperature specific heat C is analyzed in the C/T versus T^2 scale, the Schottky effect (a single-ion property in a system of non-interacting electrons) and the electron-electron interactions both yield a very similar upturn in the $T \to 0$ limit. The origin of the upturn can be unraveled from the magnetic-field dependence of the low-temperature specific heat.

1. Introduction

When studying electronic properties of rare-earth (RE) containing quasicrystals and their approximants from the low-temperature specific heat, there is a drawback which often hampers the interpretation of experimental data, namely the crystal-field (CF) splitting of the Hund's rule ground state [1]. The highly anisotropic 4f charge cloud of a RE ion (with the exception of the spherically symmetric gadolinium) interacts with the crystalline electric field and lifts the $(2J+1)$ −fold degeneracy of its ground state with the total angular momentum $\hbar\vec{J}$, producing a set of discrete energy levels. Specific heat of a system with discrete energy levels possesses an additional contribution that becomes dominant at low temperatures, known as the Schottky effect [2]. The Schottky specific heat C_{Sch} is characterized by a broad peak, appearing at a temperature that is a fraction of the energy difference $\Delta\varepsilon/k_B$ (in temperature units) between the CF-split energy levels. For small CF splittings, the peak may appear in the Kelvin- or sub-Kelvin temperature range, where it obscures the electronic specific heat. Precise determination of the electronic specific heat coefficient may thus be difficult and uncertain for quasicrystals containing RE elements. In addition, when the low-temperature total specific heat is presented in a C/T versus T^2 scale (as it is customary done), the Schottky specific heat term yields an upturn in the $T \to 0$ limit that cannot be easily distinguished from a similar upturn produced by the electron-electron interactions in exchange-enhanced systems and strongly correlated systems. The low-temperature magnetic susceptibility can also be affected by the CF effects [1,3,4]. The CF-split energy level scheme of some RE ions contains a non-magnetic ground state, whereas higher levels are magnetic. In the $T \to 0$ limit, the weight of the non-magnetic state in the electronic magnetization increases, so that the magnetic susceptibility χ decreases, resulting in a Curie-Weiss-like deviation from the linear $\chi^{-1} \propto T$ Curie behavior. However, both the CF-induced Curie-Weiss behavior of the susceptibility and the Schottky specific heat are single-ion properties, depending on the distribution of electric charges around the RE ion, but are not related to the interactions within the electronic system.

In a recent publication [5] we have demonstrated the existence of Schottky effect in the *i*-Zn-Ag-Sc-Tm Tsai-type icosahedral quasicrystal (*i*-QC) and its Zn-Sc-Tm 1/1 cubic approximant. We have first derived the CF Hamiltonian pertinent to the class of Tsai-type *i*-QCs and their approximants, which includes most of the known RE-containing quasicrystalline systems. In the Tsai-type atomic clusters, the RE elements are located on fivefold axes of a three-shell icosahedral atomic cluster that represents the backbone of the structure, so we have derived the CF Hamiltonian for the pentagonal (5-fold) rotational symmetry of the crystalline electric field. Using the leading term of this Hamiltonian, we have calculated analytically the Schottky specific heat in the presence of an external magnetic field and made comparison to the experimental specific heat of the *i*-$Zn_{74.5}Ag_{9.5}Sc_{12}Tm_4$ *i*-QC (abbreviated as *i*-Zn-Ag-Sc-Tm) and its 1/1 bcc approximant $Zn_{85.5}Sc_{11}Tm_{3.5}$ (abbreviated as 1/1 Zn-Sc-Tm). In these systems, the CF-split energy levels of thulium ions introduce Schottky contribution to the specific heat that exhibits a broad maximum at a temperature around 1 K and the maximum is shifted to higher temperatures by the application of an external magnetic field. We have also shown that even the nuclear Schottky effect, being much smaller effect than the ionic Schottky effect, is experimentally observed in the specific heat.

2. The low-temperature specific heat

The low-temperature specific heat of a RE-containing metallic alloy in the paramagnetic phase is a sum of the electronic, lattice and Schottky contributions [2], $C = \gamma T + \beta_3 T^3 + C_{Sch}$, where γ and β_3 are the electronic and lattice specific heat coefficients, respectively, and C_{Sch} is the Schottky specific heat. When analyzing the low-temperature specific heat in a C/T versus T^2 scale, one obtains for non-magnetic alloys in the absence of the Schottky contribution ($C_{Sch} = 0$) a straight line $C/T = \gamma + \beta_3 T^2$. In the presence of the Schottky contribution, the specific heat data presented in this scale show a low-temperature upturn. This is demonstrated in Fig. 1 on the example of the 1/1 Zn-Sc-Tm approximant, where the zero-field specific heat is shown in such kind of a plot. In metallic systems with incompletely filled d or f shells, electron-electron exchange interactions may provide another term to the specific heat that also produces a low-temperature upturn in the C/T versus T^2 scale. Such systems are known as the exchange-enhanced systems [6] and show instability to magnetism, i.e., they are close to a phase transition between a paramagnetic and a magnetically ordered phase. The interacting electrons give rise to long-lived spin fluctuations (similar to spin waves), which in the quantized form are called paramagnons or virtual magnons. The interaction of the electrons with paramagnons gives rise to an enhanced effective mass of the electrons (manifested in an increased γ value) and a logarithmic term in the electronic specific heat at low temperatures, which can be represented by a general equation of the form [6]

$$C = \gamma T + AT^3 \ln(T/T_{sf}) + \beta_3 T^3, \quad (1)$$

where T_{sf} is the spin fluctuation temperature. The logarithmic term is negative for $T < T_{sf}$, so that the total specific heat shows an upturn at low temperatures when plotted as C/T versus T^2. The logarithmic term in the electronic specific heat was derived also without invoking the idea of paramagnons; it was found in ^{3}He due to temperature-dependent interparticle interaction [7], suggesting that the $T^3 \ln T$ behavior of the low-temperature specific heat may be a general feature of Fermi-liquid systems.

3. Results and Conclusions

In order to demonstrate the difficulty of distinguishing between the Schottky effect and the electron-electron interactions when analyzing the low-temperature upturn in the specific heat, we present in Fig. 1 the fit of the zero-field specific heat data of the 1/1 Zn-Sc-Tm "Schottky" system with the Eq. (1) rewritten in the form $C/T = \gamma + AT^2 \ln(T/T_{sf})$, valid for an exchange-enhanced system. The fit is shown by a solid curve (the fit parameters are given in the Figure caption), where it is seen that the logarithmic term reproduces excellently the low-temperature upturn due to the Schottky effect in a limited temperature interval. The low-temperature upturn can also be reproduced qualitatively by a power-law function of a form $C/T = BT^{-n}$ (with $n > 0$), as used in the study of quantum critical phenomena arising from strongly correlated 4f electrons in Fermi-liquid systems, where the zero-field C/T diverges as a power law in the $T \to 0$ limit. The fit of the 1/1 Zn-Sc-Tm C/T data with the power-law function is shown in Fig. 1 as a dashed curve (the fit parameters are given in the Figure caption), where it is evident that it qualitatively reproduces the low-temperature upturn. The analysis of the low-temperature specific heat of RE-containing quasicrystals and approximants may thus be ambiguous, as the Schottky effect and the electron-electron interactions both yield a very similar upturn in the C/T versus T^2 scale.

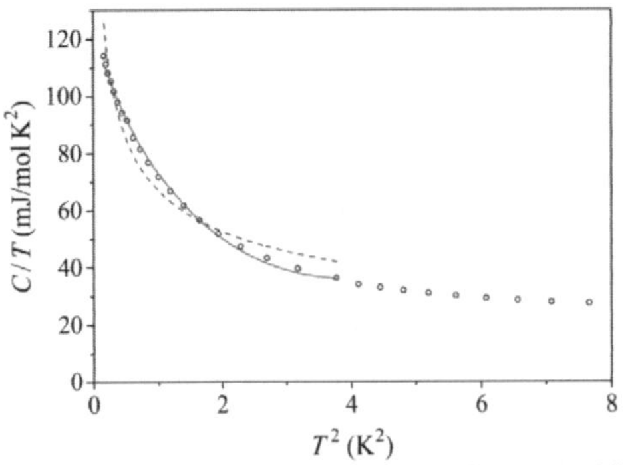

Figure 1. Low-temperature zero-field specific heat of the 1/1 Zn-Sc-Tm "Schottky" system in a C/T versus T^2 scale. Solid curve is the fit with $C/T = \gamma + AT^2 \ln(T/T_{sf})$, valid for an exchange-enhanced system (the fit parameters of the curve are γ = 127.7 mJ/mol·K^2, $A = 46$ mJ/mol·K^4, and T_{sf} = 3.3 K). Dashed curve is the fit with a power-law function $C/T = BT^{-n}$ (the fit parameters of the curve are B = 66.9 mJ/mol·K^2 (where the temperature is considered to be dimensionless) and $n = 0.7$).

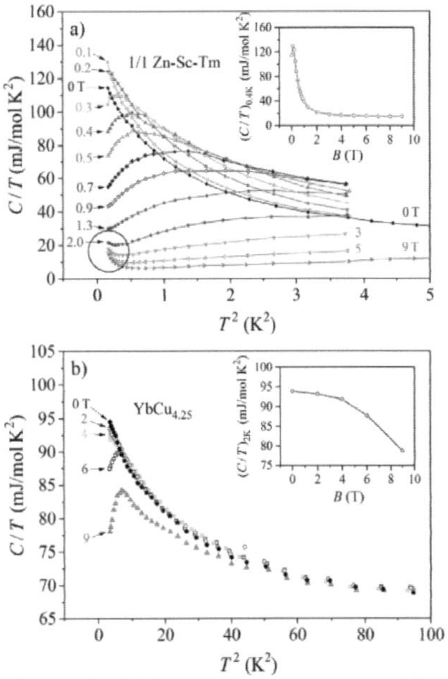

Figure 2. (a) Low-temperature specific heat of the 1/1 Zn-Sc-Tm. The upturn due to the nuclear Schottky effect is encircled. The inset shows $(C/T)_{0.4K}$ as a function of B. **(b)** Low-temperature specific heat of the $YbCu_{4.25}$ heavy-fermion compound. The inset shows $(C/T)_{2K}$ as a function of the magnetic field.

The origin of the upturn can be unraveled from the magnetic-field dependence of the low-temperature specific heat, where the Schottky effect (a single-ion property in a system of non-interacting electrons) and the electron-electron interactions in an exchange-enhanced system yield different behavior. This is demonstrated experimentally by comparing the field-dependent specific heats of the 1/1 Zn-Sc-Tm approximant and the $YbCu_{4.25}$ heavy-fermion compound, belonging to the class of strongly correlated systems. In Fig. 2, the specific heat of the 1/1 Zn-Sc-Tm "Schottky" system in the temperature interval between 0.4 and 2 K at different magnetic fields is shown in a C/T versus T^2 scale. The zero-field specific heat and the specific heat in the lowest applied field of 0.1 T show the same shape of the $T \to 0$ upturn, but the 0.1-T curve lies higher than the zero-field curve in the entire presented temperature range. The specific heat in the next higher field of 0.2 T is also larger than the zero-field specific heat, but the shape of the upturn starts to change with a tendency to round at the lowest temperature. For the fields above 0.2 T, the upturn evolves into a maximum, which moves to higher temperatures with an increasing field (a property of the Schottky maximum). The specific heat curves in the field $B \geq 0.3$ T cross the zero-field curve at some temperature, lying below the zero-field curve in the low-temperature limit and becoming higher at elevated temperatures. The Schottky maximum with its characteristic dependence on the magnetic field is thus well observed also in the C/T versus T^2 scale. In addition, the low-temperature upturn due to the nuclear Schottky effect also becomes observed in the specific heat curves at fields above 2 T (shown encircled in Fig. 2). Since the ^{169}Tm nuclei (of 100 % natural abundance) possess spin $I = 1/2$, the nuclear quadrupole interaction is zero and the nuclear Schottky effect originates from the Zeeman interaction with the effective magnetic field. Another convenient way of demonstrating the motion of the Schottky maximum with the magnetic field to higher temperatures is to plot the C/T value at the lowest investigated temperature (0.4 K in our case) as a function of the magnetic field. This is presented in the inset of Fig. 2, where the $(C/T)_{0.4K}$ value first increases with the magnetic field, passes through a maximum and then decreases at higher fields, following the field-induced shift of the Schottky maximum to higher temperatures. The asymptotic $(C/T)_{0.4K}$ value reached in the high-field limit is close to zero and is determined by the nuclear Schottky specific heat. The magnetic-field dependence of the low-temperature specific heat of an exchange-enhanced system is different, as will be illustrated on the example of the $YbCu_{4.25}$ heavy-fermion (HF) compound [8]. $YbCu_{4.25}$ is a giant-unit-cell complex intermetallic, which was reported to be a "moderate" HF compound with the increase of the thermal effective mass of the charge carriers by a factor of 84.5 with respect to the free-electron mass. The low-temperature specific heat of the $YbCu_{4.25}$ in magnetic fields between zero and 9 T is shown in Fig. 2 in a C/T versus T^2 scale. The zero-field specific heat shows a $T^3 \ln T$ logarithmic upturn, present in this compound due to the strong exchange interactions between the Yb f-electrons and the conduction electrons. In a magnetic field, the specific heat at low temperatures is lowered with respect to the zero-field curve for any field value, and the decrease is stronger in higher fields. In addition, a high-enough magnetic field (above 6 T in this case) introduces sharp maximum in the C/T curve at a temperature, which depends weakly on the magnetic field. When plotting the $(C/T)_{2K}$ value (2 K was the lowest investigated temperature) as a function of the magnetic field (inset in Fig. 2), the curve does not exhibit a maximum as in the case of the Schottky effect, but decreases monotonously with the field. The field-induced decrease of the $(C/T)_{2K}$ value is relatively small; in the 9-T field it has dropped only by 17 % with respect to the zero-field value (note the highly expanded vertical scale in the inset of Fig. 2). This behavior is typical for HF systems, where it originates from the fact that the external magnetic field gradually destroys the Kondo-type compensation of the f-electron magnetic moments by the oppositely spin-polarized conduction-electron cloud that builds up around the RE ion, making the f-electron moments to reappear in the field. Therefore, careful inspection of the low-temperature specific heat behavior in a magnetic field is crucial to distinguish whether the upturn originates from the Schottky effect or it is due to the electron-electron interactions in an exchange-enhanced system.

References

[1] Fulde P 1978 *Crystal Fields* (in *Handbook on the Physics and Chemistry of Rare Earths*) Gschneidner K A Jr and Eyring L eds (Amsterdam: North-Holland) Ch 17
[2] Tari A 2003 *The Specific Heat of Matter at Low Temperatures* (London: Imperial College Press) Ch 6
[3] Gignoux D, Givord F and Lemaire R 1975 *Phys. Rev.* B **12** 3878
[4] Vogt O and Cooper B R 1968 *J. Appl. Phys.* **39** 1202
[5] Jazbec S *et al* 2016 *Phys. Rev.* B **93** 054208
[6] Ikeda K *et al* 1991 *J. Magn. Magn. Mater.* **100** 292
[7] Tripathi D N and Mishra S 1996 *Physica* B **226** 338
[8] Dolinšek J *et al* 2010 Solid State Commun **150** 1629

Magnetic Properties of the Au-Al-Yb Approximant under Hydrostatic Pressure

Shuya Matsukawa[1], Kazuhiko Deguchi[1], Keiichiro Imura[1], Tsutomu Ishimasa[2], and Noriaki K. Sato[1]

[1]Department of Physics, Graduate School of Science, Nagoya University, Nagoya 464-8602, Japan

[2]Division of Applied Physics, Graduate School of Engineering, Hokkaido University, Sapporo 060-8628, Japan

E-mail: matsukawa.shuya@h.mbox.nagoya-u.ac.jp

Abstract. We report on ac magnetic susceptibility measurements of the Au-Al-Yb approximant under pressure. It is known that the approximant shows unconventional quantum criticality at the critical pressure $P_c \sim 2$ GPa, and further that the magnetic ordered state emerges at a low temperature ($T_g \sim 100$ mK) for $P > P_c$. We find that the nonlinear susceptibility shows a peak around the temperature T_g at which the linear susceptibility peaks. We also find that T_g increases with frequency of the modulation field. From these results, we suggest that the magnetic ordered state of the approximant below T_g is a spin-glass-like short-range ordered state.

1. Introduction

Quantum critical phenomena have attracted much interest in last decades. These are observed near the quantum critical point (QCP) of strongly correlated materials where the second-order transition temperature vanishes to zero. The low-temperature properties of these materials differ from those of ordinary Fermi systems. Recently, the unconventional quantum critical phenomenon was observed in the Au-Al-Yb icosahedral quasicrystal: As the temperature T is decreased toward zero temperature, the physical properties diverge like $\chi \propto T^{-0.51}$ and $C/T \propto -\ln T$ (where χ is the magnetic susceptibility and C is the specific heat), suggesting that the system is just at the QCP without tuning [1]. These critical indices are very unusual and similar to those of the heavy fermion crystals β-$YbAlB_4$ and $YbRh_2Si_2$ [2]-[5]. In addition, the quantum critical behavior of the quasicrystal is robust against the application of hydrostatic pressure: At 1.54 GPa, the divergence of the magnetic susceptibility with the same critical index ($\gamma \sim 0.5$) was observed [1]. In contrast, the Au-Al-Yb approximant does not show such the diverging behavior at ambient pressure: The magnetic susceptibility is described as $1/\chi \propto T^{0.5}+$ constant [6]. Here we note that the presence of the constant term means that χ does not diverge (and instead saturates toward this constant susceptibility) as $T \to 0$, implying that the system is away from the QCP. At the critical pressure $P_c \sim 2$ GPa, however, the constant term vanishes to zero and χ diverges with the same critical index as the quasicrystal [6]. By comparing the critical behaviors with theories, the quantum criticality is likely ascribe to the critical valence fluctuation [6],[7]. At high pressures greater than P_c, the χ of the approximant shows a peak anomaly at a very low temperature T_g ($\sim$ 100 mK) [6]. It is unclear whether the magnetic ordered state below T_g is an antiferomagnetically long-range ordered state or a spin-glass-like short-range ordered state. To elucidate the long-range or short-range ordered state, it is useful to measure the nonlinear susceptibility χ_2 defined as follows,

$$\chi(H) = \frac{dM}{dH} = \chi_0 + \chi_2 H^2 + \cdots. \quad (1)$$

Here, M is a magnetization, and the χ_0 and χ_2 are the linear and nonlinear susceptibility, respectively. Figures 1(a) and (b) show the temperature dependence of χ_2 around an antiferromagnetic ordering temperature T_N (a) and a spin-glass freezing temperature T_g (b) [8]. In the case of the antiferromagnetic order, -χ_2 shows no divergence and a sign change at T_N. In the case of the spin-glass transition, on the other hand, -χ_2 diverges at both sides of T_g. To determine the long-range or short-range ordering, it is also useful to measure the modulation frequency dependence of the transition temperature [9]. In the case of an antiferromagnetic order, T_N is independent of frequency of the modulation field, while in the spin-glass order, T_g increases with frequency. Here, we report the nonlinear magnetic susceptibility and the modulation-field frequency dependence of the transition temperature at high pressure.

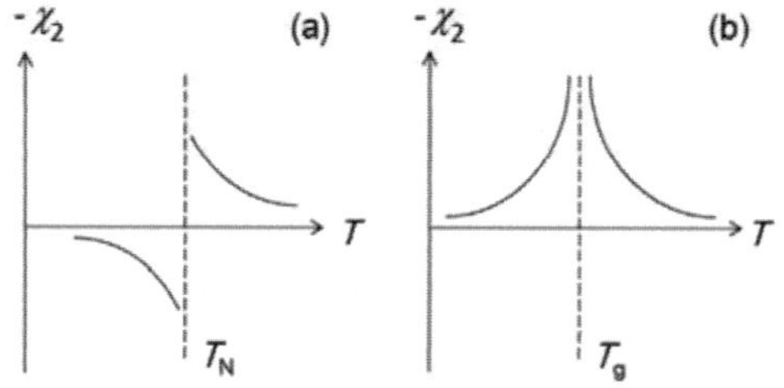

Figure 1. Schematic illustration of the nonlinear susceptibility of an antiferromagnetic order (a) and a spin-glass-like short-range order (b) as a function of temperature. T_N and T_g indicate a Néel temperature and a freezing temperature, respectively.

2. Experimental details

The approximant crystal with the nominal composition of $Au_{49}Al_{36}Yb_{15}$ was prepared by arc-melting starting materials of 4N (99.99 % pure)-Au, 5N-Al, and 3N-Yb and subsequently annealing the obtained alloy-ingot in an evacuated quartz ampoule at 750 ℃ for 116 hours [10]. The ac magnetic susceptibility χ was measured using the conventional mutual inductance method. A modulation field (frequency in the range of 10.03 to 3003 Hz and amplitude of 0.1 Oe) was superimposed on a dc magnetic field H supplied by a superconducting magnet. To calibrate the ac magnetic susceptibility, we measured the dc magnetization M using a MPMS machine (Quantum Design) at pressures of up to 1.2 GPa, above which we extrapolated the dc magnetization data. Hydrostatic pressure was generated using a NiCrAl-BeCu piston cylinder cell for the measurement of the ac magnetic susceptibility, together with Daphne oil 7373 as a pressure-transmitting medium. The pressure at low temperature was determined from the superconducting transition temperature of indium [11] that was put into the pressure cell together with the sample.

3. Results and discussion

Figure 2 shows the temperature dependence of the ac magnetic susceptibility $\chi(T)$ at P = 2.79 GPa for H = 0. There appears an anomaly suggestive of a transition into either an

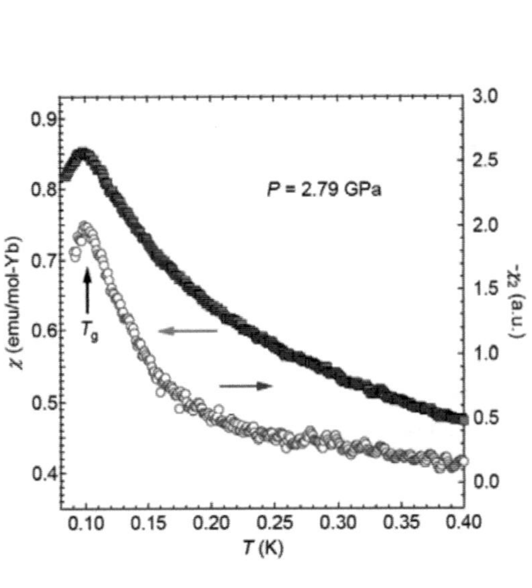

Figure 2. (Color online) Temperature dependence of the conventional susceptibility, χ $(= \chi_0)$, (left axis) and the nonlinear susceptibility, -χ_2, (right axis) of the $Au_{49}Al_{36}Yb_{15}$ approximant at $P = 2.79$ GPa.

antiferromagnetically long-range-ordered state or a spin-glass-like short-range-ordered state. The nonlinear susceptibility χ_2 is evaluated as the initial slope of the χ vs H^2 curve. We assumed here that the χ is proportional to H^2 below 300 Oe, and estimated χ_2 using the following equation,

$$\chi_2(T) = \frac{\chi(T, H) - \chi(T, 0)}{H^2}. \tag{2}$$

Thus obtained results are also plotted in Fig. 2, together with χ_0. Note that $-\chi_2$ forms a positive peak around T_g at which the linear susceptibility peaks. This strongly suggests that the ordering is of short-range nature, which is compatible with the fact that there are only limited number of examples of the long-range ordered icosahedral-approximant crystal [12].

Figure 3 shows the temperature dependence of the ac magnetic susceptibility measured at several frequencies of the modulation field. We clearly observe a peak at low temperature although it is not obvious for 3003 Hz. The peak slightly but definitely shifts to higher temperatures as the frequency increases. This results are fit to the following equation (Vogel-Fulcher law),

$$f = f_0 \exp\left[-\frac{W}{k_B(T_g - T_0)}\right]. \tag{3}$$

Here, f_0, W, and k_B are the characteristic frequency, the activation energy, and the Boltzmann constant, respectively. T_0 is called the 'ideal glass' temperature although its physical meaning is unclear. From the results shown in Fig. 4, we obtain the following values: $f_0 = 10^{13}$ Hz, $W/k_B = 760 \pm 100$ mK, and $T_0 = 68 \pm 4$ mK. This finding gives a supporting evidence for the glass-like transition at T_g.

There are two possible origins of the quantum criticality in the approximant. One is due to the QCP of the glass transition. However, this possibility is presumably excluded because the uniform susceptibility will not diverge at $T_g = 0$. The second possibility is due to the critical valence fluctuation. As mentioned in the introduction part, this possibility remains most probable. For the quasicrystal, we have detected no magnetic ordering as observed in the

IOP Conf. Series: Journal of Physics: Conf. Series **809** (2017) 012013 doi:10.1088/1742-6596/809/1/012013

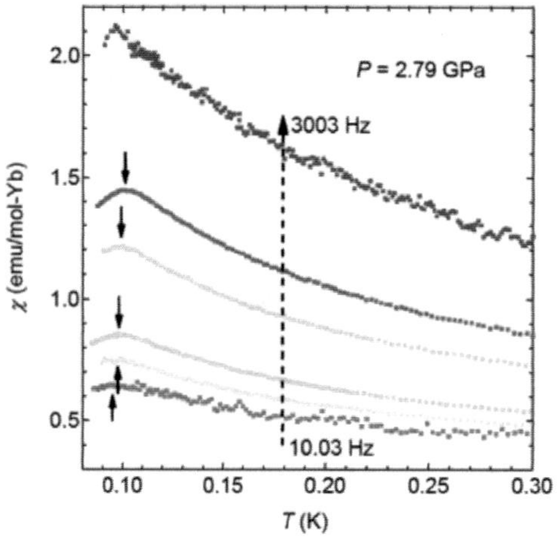

Figure 3. (Color online) Temperature dependence of the magnetic susceptibility of the Au-Al-Yb approximant at $P = 2.79$ GPa, measured at different applied frequencies $f = 10.03$, 50.03, 100.3, 500.3, 1003, and 3003 Hz.

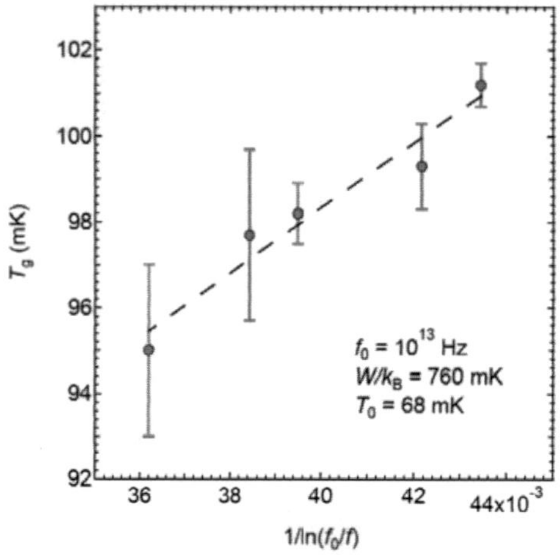

Figure 4. (Color online) The variation of the freezing temperature T_g with $1/\ln(f_0/f)$, using $f_0 = 10^{13}$ Hz. The dash line is a guide to the eyes.

approximant. Therefore, the first possibility mentioned above is unambiguously excluded. The second possibility remains to be a possible origin.

4. Summary

We measured the ac magnetic susceptibility of the Au-Al-Yb approximant crystal down to about 80 mK under hydrostatic pressure. We found that the nonlinear magnetic susceptibility at 2.79 GPa shows a peak anomaly at $T_g \sim 100$ mK and T_g increases with frequency of the modulation field, which results in the Vogel-Fulcher law. These results suggest that the magnetic ordered state of the Au-Al-Yb approximant below T_g is a spin-glass-like short-range ordered state.

5. Acknowledgments

The authors thank Yukinori Tanaka and Shin Yamamoto for support of the experiments. This work was financially supported by a Grant-in-Aid for Scientific Research from JSPS, KAKENHI (Nos. 26610100, 15H02111, and 15H03685) and Program for Leading Graduate Schools "Integrative Graduate Education and Research Program in Green Natural Sciences", MEXT, Japan.

6. References

[1] Deguchi K, Matsukawa S, Sato N K, Hattori T, Ishida K, Takakura H, and Ishimasa T 2012 *Nat. Mater.* **11** 1013
[2] Nakatsuji S, Kuga K, Machida Y, Tayama T, Sakakibara T, Karaki Y, Ishimoto H, Yonezawa S, Maeno Y, Pearson E, Lonzarich G G, Balicas L, Lee H, and Fisk Z 2008 *Nature Phys.* **4** 603
[3] Matsumoto Y, Nakatsuji S, Kuga K, Karaki Y, Horie N, Shimura Y, Sakakibara T, Nevidomskyy A H, and Coleman P 2011 *Science* **331** 316
[4] Trovarelli O, Geibel C, Mederle S, Langhammer C, Grosche F M, Gegenwart P, Lang M, Sparn G, and Steglich F 2000 *Phys. Rev. Lett.* **85** 626
[5] Gegenwart P, Westerkamp T, Krellner C, Tokiwa Y, Paschen S, Geibel C, Steglich F, Abrahams E, and Si Q 2007 *Science* **16** 969
[6] Matsukawa S, Deguchi K, Imura K, Ishimasa T, and Sato N K 2016 *J. Phys. Soc. Jpn.* **85** 063706
[7] Watanabe S and Miyake K 2010 *Phys. Rev. Lett.* **105** 186403
[8] Fujiki S, Katsuwa S 1981 *Prog. Theor. Phys.* **65** 1130
[9] Mydosh J A 1993 *Spin Glasses : An Experimental Introduction* (London: Taylor & Francis)
[10] Ishimasa T, Tanaka Y, and Kashimoto S 2011 *Phil. Mag.* **91** 4218
[11] Jennings L D, and Swenson C A 1958 *Phys. Rev.* **112** 31
[12] Nakayama M, Tanaka K, Matsukawa S, Deguchi K, Imura K, Takakura H, Ishimasa T, and Sato N K 2015 *J. Phys. Soc. Jpn.* **84** 024721

Unified cluster-based description of valence bands in AlIr, $RuAl_2$, $RuGa_3$ and Al–TM quasicrystalline approximants

K Kitahara[1], Y Takagiwa[1,2] and K Kimura[1]
[1] Department of Advanced Materials Science, The University of Tokyo, Kashiwanoha, Kashiwa 277-8561, Japan
[2] National Institute for Materials Science, Sengen, Tsukuba 305-0047, Japan
E-mail: kitahara@phys.mm.t.u-tokyo.ac.jp

Abstract. Formation of the valence bands in AlIr, $RuAl_2$ and $RuGa_3$ was analysed on the basis of Wannier functions. It was sufficient to consider nine (*s*-, *p*- and *d*-like) Wannier functions per *cluster* centred on each transition metal for describing the valence bands. As in the case of Al–Cu–Ir quasicrystalline approximant, which we reported previously [Kitahara K, Takagiwa Y and Kimura K 2015 *J. Phys. Soc. Jpn.* **84** 014703-1–8], covalent bonds between clusters form where the distance between neighbouring clusters is about 0.3 nm, making conduction states as antibonding functions. sp^3d^2-, sp^3- and p_y-like Wannier functions are used for the covalent bonds in AlIr, $RuAl_2$ and $RuGa_3$, respectively.

1. Introduction

Formation of pseudogap or narrow bandgap in some of group 13 element–transition metal (TM) intermetallic compounds including Al–TM quasicrystalline approximants had extensively been studied, and importance of hybridisation between *sp*-functions of group 13 elements (or free electrons in jellium) and *d*-functions of TM atoms were realised [1]. However the concept of *sp*–*d* hybridisation cannot completely explain the formation of bandgap, especially the number of valence bands. For example, in the case of $RuAl_2$, there are seven valence bands per ruthenium atom without considering spin degeneracy. While five out of the seven bands may be associated with *sp*–*d* bonding functions, the origin of the other two bands is unclear.

We studied the electronic band structures in AlIr (CsCl-type), $RuAl_2$ ($TiSi_2$-type), $RuGa_3$ ($CoGa_3$-type) and Al–Cu–Ir quasicrystalline approximant as typical intermetallic compounds with pseudogap or narrow bandgap. To analyse the band structures, we used maximally localised Wannier functions [2], which are representative functions corresponding to the bands in a specified energy range. We have already reported analysis on the Al–Cu–Ir approximant [3], where we described the valence and a part of the conduction bands on the basis of Wannier functions centred on *clusters* of 0.2 nm radius, and the number of valence bands (173) was decomposed in terms of the numbers of TMs ($t = 23$), clusters ($c = 16$) and covalent bonds between the clusters ($b = 6$) per primitive unit cell as $5t + 4c - b$ without considering spin degeneracy. As we shall show, fundamentally the same description can be applicable to the other three intermetallic compounds, AlIr, $RuAl_2$ and $RuGa_3$.

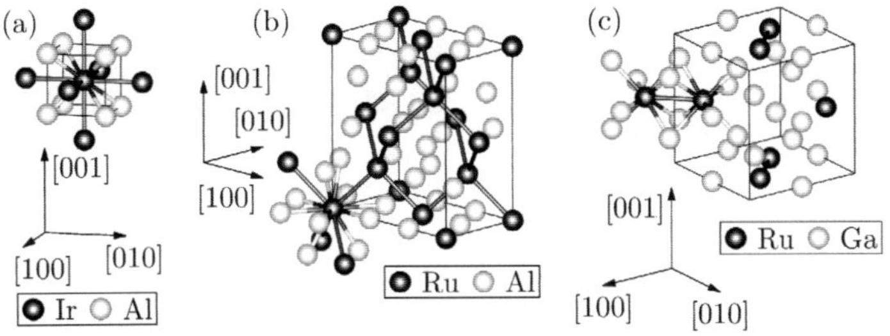

Figure 1. Crystalline structures of (a) AlIr, (b) $RuAl_2$ and (c) $RuGa_3$.

2. Methods

Self-consistent ground-state electronic structure calculations were performed using an all-electron full-potential linearised augmented plane-wave code [4] within the local density approximation of the density functional theory [5] for AlIr [CsCl-type, space group $Pm\bar{3}m$, Figure 1(a)], $RuAl_2$ [$TiSi_2$-type, space group $Fddd$, Figure 1(b)] and $RuGa_3$ [$CoGa_3$-type, space group $P4_2/mnm$, Figure 1(c)]. We used experimental lattice constants and atomic positions for AlIr [6], $RuAl_2$ [7] and $RuGa_3$ [8]. Spin polarisation and spin–orbit coupling were not considered in this study. Unshifted $16 \times 16 \times 16$, $8 \times 8 \times 8$ and $4 \times 4 \times 4$ wave-vector grids were used to construct Wannier functions for AlIr, $RuAl_2$ and $RuGa_3$, respectively. We constructed sets of Wannier functions which span the valence and a part of the conduction bands and which are localised around TMs in similar ways as described in [3].

3. Results

We constructed nine Wannier functions per iridium atom for AlIr. As shown in Figure 2(a), these Wannier functions contain all the valence states. While three t_{2g}-like functions are distributed within the valence bands [Figure 2(b)], the other six (s-, e_g- and p-like) functions significantly contain conduction states [Figures 2(c)–2(e)]. To analyse how these functions split into valence and conduction states, we constructed sp^3d^2-like hybrid functions oriented toward $\langle 100 \rangle$ directions as linear combinations of the six functions [Figures 2(f) and 2(i)]. Then, we solved partial Hamiltonians between pairs of sp^3d^2-like functions and obtained pairs of bonding [Figure 2(j)] and antibonding [Figure 2(k)] functions. As shown in Figures 2(g) and 2(h), the bonding and antibonding functions were mainly composed of valence and conduction states, respectively.

We constructed nine Wannier functions per ruthenium atom for each of $RuAl_2$ and $RuGa_3$. These Wannier functions completely describe the valence bands in $RuAl_2$ and $RuGa_3$, and the nine functions can be classified into one s-like, five d-like and three p-like functions or, strictly speaking, linear combinations of them.

In the case of $RuAl_2$, $d_{3y^2-r^2}$-like, d_{zx}-like, p_x-like and p_z-like functions significantly contain conduction states, and hence we constructed sp^3-like hybrid functions oriented toward $\langle 111 \rangle$ directions—we call the functions sp^3 instead of p^2d^2 for simplicity as the symmetry is the same, i.e. both belong to the same reducible representation under the point group of the ruthenium site—as linear combinations of the four functions [Figure 3(a)]. Bonding [Figure 3(b)] and antibonding [Figure 3(c)] functions between two sp^3-like functions at neighbouring ruthenium sites were mainly composed of valence and conduction states, respectively [see Supplementary Figures S1(a)–S1(m) for the partial densities of states projected on each Wannier function].

In the case of $RuGa_3$, only p_y-like function [Figure 4(a)] contains conduction states. Bonding [Figure 4(b)] and antibonding [Figure 4(c)] functions between two p_y-like functions at neighbouring ruthenium sites were mainly composed of valence and conduction states, respectively [see Supplementary Figures S2(a)–S2(l) for the partial densities of states projected on each Wannier function].

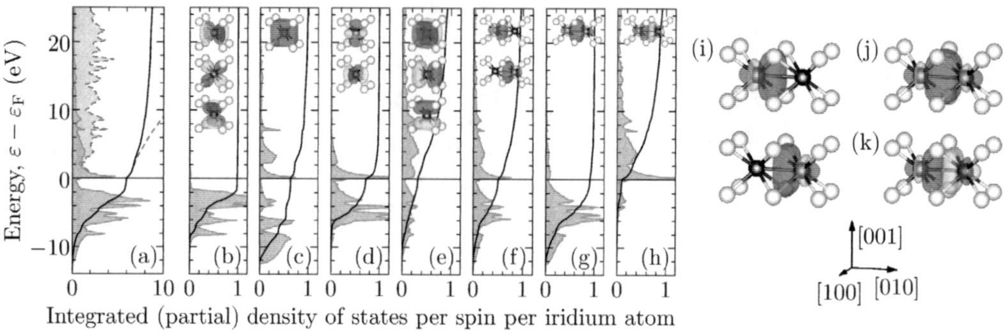

Figure 2. (colour online) Integrated partial densities of states in AlIr projected on (a) all the Wannier functions, (b) t_{2g}-like, (c) s-like, (d) e_g-like, (e) p-like, (f) sp^3d^2-like hybrid, (g) sp^3d^2–sp^3d^2 bonding and (h) sp^3d^2–sp^3d^2 antibonding functions (solid lines, black). In each graph, partial density of states is shown in an arbitrary unit (filled area, light red), and the corresponding functions are visualised (insets). In Figure (a), integrated *total* density of states (dashed line, grey) and corresponding density of states (in an arbitrary unit, filled area, light grey) are also shown. Figures (i), (j) and (k) show magnified images of a pair of sp^3d^2-like hybrid, an sp^3d^2–sp^3d^2 bonding and an sp^3d^2–sp^3d^2 antibonding functions, respectively.

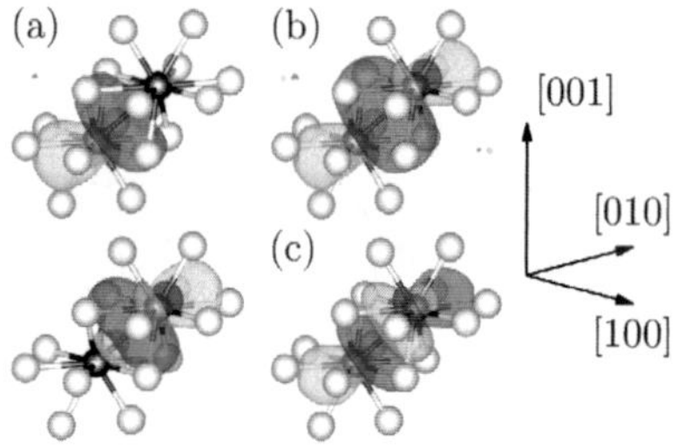

Figure 3. (colour online) Visualised images of (a) a pair of sp^3-like hybrid, (b) an sp^3–sp^3 bonding and (c) an sp^3–sp^3 antibonding functions in $RuAl_2$.

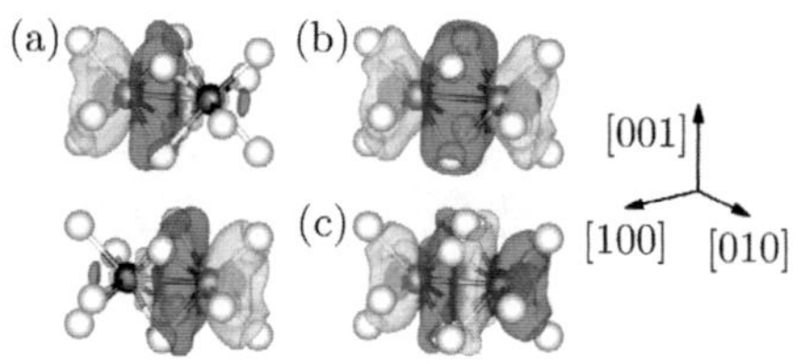

Figure 4. (colour online) Visualised images of (a) a pair of p_y-like, (b) a p_y–p_y bonding and (c) a p_y–p_y antibonding functions in $RuGa_3$.

4. Discussion

To summarise our results, it is convenient to consider clusters of which centres are the same as those of the constructed Wannier functions. In the cases of AlIr, $RuAl_2$ and $RuGa_3$, the centres of such clusters are the sites of TMs; therefore, the numbers of TMs (t) and clusters (c) per primitive unit cell are $t = c = 1$ for AlIr, $t = c = 2$ for $RuAl_2$ and $t = c = 4$ for $RuGa_3$. For each cluster, we constructed one s-, three p- and five d-like Wannier functions as candidates of valence states. It corresponds to a part of our counting rule $5t + 4c$, i.e. five d-like functions per TM and one s- and three p-like functions per cluster. Formation of pairs of bonding and antibonding functions between Wannier functions at neighbouring sites can be considered as covalent bonds between the clusters. Such covalent bonds form where the distance between two clusters is about 0.3 nm as discussed in [3]. Six sp^3d^2-, four sp^3- and one p_y-like functions per cluster are used for the covalent bonds in AlIr, $RuAl_2$ and $RuGa_3$, respectively, making a conduction state as an antibonding function per pair of clusters. Thus, the number of covalent bonds between the clusters (b) per primitive unit cell is $b = 3$ for AlIr, $b = 4$ for $RuAl_2$ and $b = 2$ for $RuGa_3$, and then the number of valence bands is correctly evaluated on the basis of $5t + 4c - b$ rule as

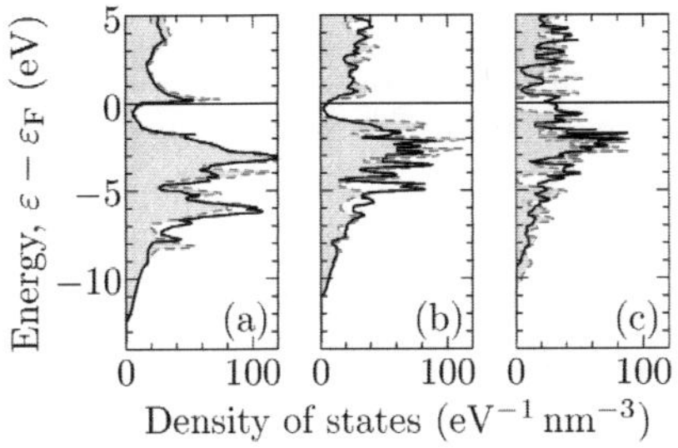

Figure 5. Densities of states in approximated *TMs-in-jellium* (solid lines, black) and original (filled areas, light grey) models of (a) AlIr, (b) $RuAl_2$ and (c) $RuGa_3$.

$5 \times 1 + 4 \times 1 - 3 = 6$ for AlIr, $5 \times 2 + 4 \times 2 - 4 = 14$ for $RuAl_2$ and $5 \times 4 + 4 \times 4 - 2 = 34$ for $RuGa_3$.

We should note that, Yannello and Fredrickson proposed $18 - n$ rule for the number of valence electrons per TM in TM–main group intermetallic phases [9], where n is the number of electrons in bonding functions between TMs. This rule is a special case of our $5t + 4c - b$ rule, which reduces to $18 - n$ rule when $t = c$; in such case, $2b = tn$, and therefore $2(5t + 4t - b) = t(18 - n)$. In the present cases, we do not need to consider TMs and clusters separately, and therefore $18 - n$ rule is sufficient. However, in some cases (at least in the case of Al–Cu–Ir quasicrystalline approximant [3]), the numbers of TMs and clusters can differ, and $5t + 4c - b$ rule is required.

To associate all the valence states with TMs implies that formal charges of iridium in AlIr, ruthenium in $RuAl_2$ and ruthenium in $RuGa_3$ are −3, −6 and −9, respectively. Öğüt and Rabe [10] calculated the Born effective charge tensor of ruthenium in $RuAl_2$, and the directional average of the tensor was about −6.2. This fact supports the formal charge of ruthenium in $RuAl_2$. To examine the role of aluminium or gallium in formation of the electronic structures, we also calculated the electronic band structures in *TMs-in-jellium* models, where each aluminium or gallium in AlIr, $RuAl_2$ and $RuGa_3$ are replaced with homogeneous charges of 3+ and three free electrons. In Figures 5(a)–(c), the densities of states in these approximated models are compared with those of the original models. In the cases of AlIr and $RuAl_2$, the densities of states of both the original and the approximated models are almost the same. Even in the case of $RuGa_3$, where 75% of atoms are replaced with jellium in the approximated model, overall structure of the density of states is not drastically changed, and gap-like structure remains around 1 eV above the Fermi energy. It indicates that the role of aluminium or gallium is, as a first approximation, just to offer free electrons to the system.

Finally, as our description is based on a short-range mechanism, it is very interesting whether it can be applicable to quasicrystals or higher order approximants, too; if not, there is a room to consider a long-range mechanism.

Acknowledgments
This work was supported by JSPS KAKENHI Grant Number JP16H04489.

References
[1] Trambly de Laissardière G, Nguyen-Manh D and Mayou D 2005 *Prog. Mater. Sci.* **50** 679–788
[2] Marzari N, Mostofi A A, Yates J R, Souza I and Vanderbilt D 2012 *Rev. Mod. Phys.* **84** 1419–75
[3] Kitahara K, Takagiwa Y and Kimura K 2015 *J. Phys. Soc. Jpn.* **84** 014703-1–8
[4] A code developed on the basis of Elk code (elk.sourceforge.net) for our purposes
[5] Perdew J P and Wang Y 1992 *Phys. Rev.* B **45** 13244–9
[6] Schulz V H, Ritapal K, Bronger W and Klemm W 1968 *Z. Anorg. Allg. Chem.* **357** 299–313
[7] Edshammar L-E 1966 *Acta Chem. Scand.* **20** 427–31
[8] Häussermann U, Boström M, Viklund P, Rapp Ö and Björnängen T 2002 *J. Solid State Chem.* **165** 94–9
[9] Yannello V J and Fredrickson D C 2015 *Inorg. Chem.* **54** 11385–98
[10] Öğüt S and Rabe K M 1996 *Phys. Rev.* B **54** R8297–300

Aging effects of dodecagonal quasicrystal formed in Mn-Cr-Ni-Si alloys

Kotoba Toyonaga and Tsutomu Ishimasa

Graduate School of Engineering, Hokkaido University, Kita-ku, 060-8628 Sapporo, Japan

E-mail: ishimasa@eng.hokudai.ac.jp

Abstract The formation of Mn-Cr-Ni-Si dodecagonal quasicrystal has been studied at 600 and 700 °C. The growth process of the quasicrystal from as-cast β-Mn phase was explained by Johnson-Mehl-Avrami equation. In order to evaluate structural quality of the quasicrystal, distortion in the diffraction pattern observed along the 12-fold axis was analyzed. This analysis indicated that the aging at 600 °C for 100 h is the best condition to synthesize the quasicrystal.

1. Introduction

A dodecagonal quasicrystal is a three-dimensional structure that is characterized by a unique 12-fold symmetry axis in the reciprocal space. Recently a new dodecagonal quasicrystal has been found in Mn-Cr-Ni-Si system, of which five-dimensional space group is $P12_6/mmc$ [1]. High resolution electron microscopy has revealed that the projection of the quasicrystal along the 12-fold axis is regarded as nonperiodic arrangement of a square and an equilateral triangle with the common edge length a = 4.560 Å [2]. Furthermore, geometrical analysis in the phason space has revealed the presence of an acceptance region that indicates the quasiperiodic order in a limited region within approximately 130 Å in diameter. This quasicrystal is precipitated from a β-Mn phase by heat treatment at 700 °C [1]. The purpose of the present study is to clarify aging effects on the formation of the quasicrystal.

2. Experimental procedures

At the composition $Mn_{72.0}Cr_{5.5}Ni_{5.0}Si_{17.5}$, alloy specimens were prepared using an arc furnace in an Ar atmosphere. The as-cast alloy just after melting included the cubic β-Mn type phase exclusively. They were further aged in an electric furnace at 600 or 700 °C for 57 - 611 h. The specimen enclosed in a silica ampoule was put into the furnace already set at each aging temperature beforehand. This procedure enables us to study the effect of the aging. Structural characterization was carried out using both a transmission electron microscope JEM200CS operated at 200 kV and a powder X-ray diffractometer RINT2200 using Cu Kα-radiation. In order to quantify both integral intensity and width of reflections, profile decomposition was carried out assuming pseudo-Voigt function for each reflection, where the contributions of $K\alpha_1$ and $K\alpha_2$ components were considered. Selected-area diffraction patterns were observed along the 12-fold axis using an aperture corresponding to approximately 2500 Å in diameter. The indexing scheme of a dodecagonal quasicrystal proposed in [3] is used throughout this report.

3. Results and Discussions

Figure 1 presents a powder X-ray diffraction pattern of the $Mn_{72.0}Cr_{5.5}Ni_{5.0}Si_{17.5}$ alloy that was aged at 600 °C for 611 h. The dodecagonal quasicrystal coexists with the dominant cubic β-Mn type phase with the lattice parameter 6.270 (1) Å. The dodecagonal quasicrystal exhibits broad peaks as indicated bars at the lower part in Figure 1a. The lattice parameters were estimated by modified Cohen's method to be a = 4.558 (3) Å and c = 4.620 (4) Å. Here the parameters a and c correspond to the common edge length of the tiles and the period along the 12-fold axis, respectively. In Figure 1b, result of peak decomposition is presented, in which there are four reflections of the dodecagonal quasicrystal.

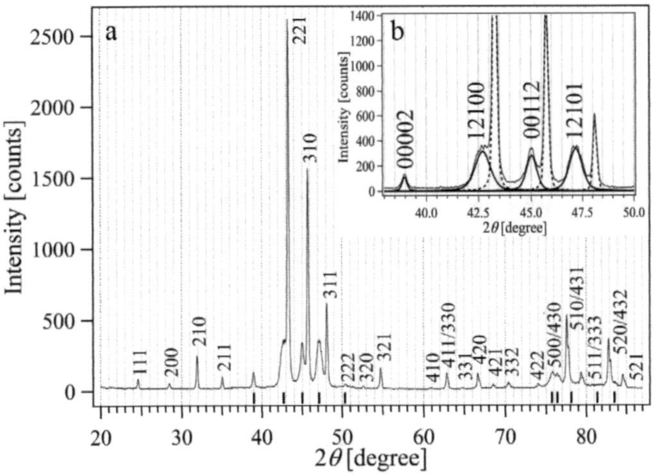

Figure 1. (a) Powder X-ray diffraction pattern of $Mn_{72.0}Cr_{5.5}Ni_{5.0}Si_{17.5}$. Indices of the β-Mn type phase are inserted. (b) Magnified part of the X-ray diffraction pattern. Four reflections can be detected: 00002, 12100, 00112, and 12101.

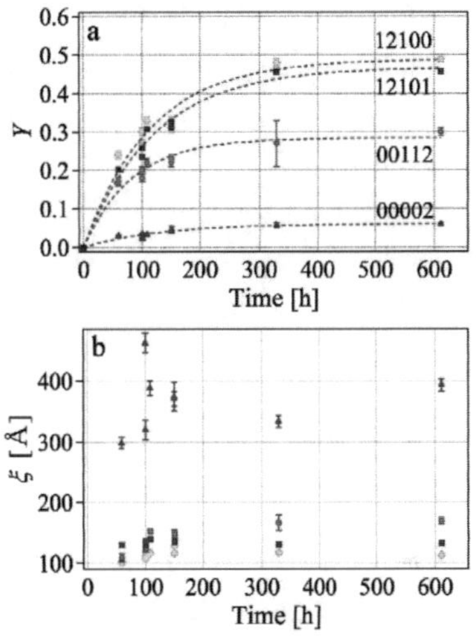

Figure 2. Effect of aging at 600 °C. (a) Normalized intensity, (b) Correlation length determined by the four reflections.

Figure 2 shows the growth of the quasicrystal during the aging at 600 °C. In Figure 2a, normalized intensities, $Y = I_g/I_{221}$, of the following four reflections are presented: $\boldsymbol{g}$ = 00002, 12100, 00112, 12101. For the normalization, the intensity of the 221 reflection of the β-Mn type phase was used as a standard. Strictly speaking, the amount of the β-Mn type phase was decreased during the aging, but the decrease was small enough for this analysis. The growth process is well explained by Johnson-Mehl-Avrami Equation: $Y = Y_0(1 - \exp(-Kt^n))$. Here t [h] is aging time, and Y_0 is constant. K is the rate parameter. When the exponent n = 1 was used, the excellent agreement shown in Figure 2a was obtained. The parameter K is approximately 0.010 h^{-1} for all the reflections. Similar growth was observed also for aging temperature 700 °C with larger K, approximately 0.020 h^{-1}. These results indicate gradual growth of the quasicrystal during the aging, and suggest stability or metastability of the quasicrystal at these temperatures. Figure 2b presents evolution of the correlation length ξ [Å] determined from the peak width $\Delta\theta$, full width at half maxima, using the following equation: $\xi = \lambda/2\Delta\theta\cos\theta$. Here λ and θ denote the wavelength of the X-ray and the Bragg angle, respectively.

Correlation lengths were slightly increased during the period between 50 and 150 h, and reached to each final value. The correlation length along the quasiperiodic direction is approximately 100 Å, but that along the periodic direction is much larger. (Notice the difference between 12100 and 00002 reflections.) Similar behavior was also observed at 700 °C. The correlation length along the quasiperiodic direction agrees with the domain size observed in the electron micrograph [2].

Examples of electron diffraction patterns are presented in Figure 3. The 12-fold symmetric pattern in Figure 3a was observed in the specimen A aged at 600 °C for 100 h, where the suffix "A" denotes one of two specimens synthesized under the same aging condition. All the reflections are indexed by the indexing scheme with $h_5 = 0$ [3], where the lattice parameters are a = 4.558 Å and c = 4.620 Å. Both spot-like shapes of reflections and weak diffuse scattering indicate high structural quality of this quasicrystal. However, deviations from the ideal pattern were frequently observed. In Figure 3b, there is diffuse scattering. The pattern in Figure 3c shows 6-fold like symmetry rather than 12-fold.

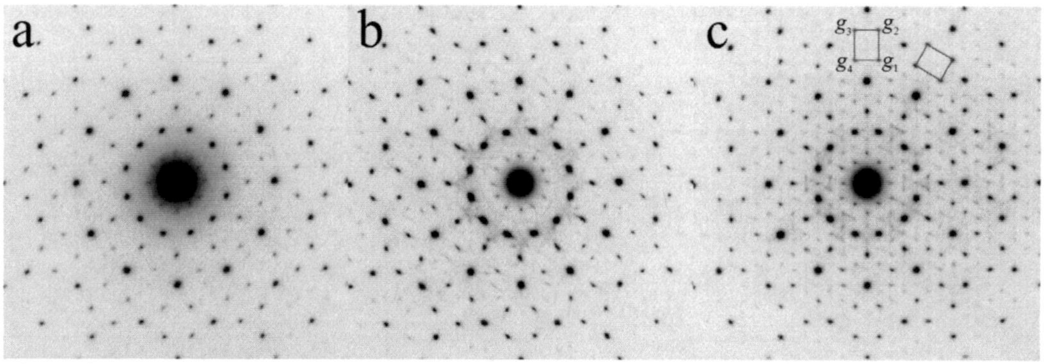

Figure 3. Electron diffraction patterns. (a) Nearly perfect pattern with small standard deviations, $\sigma(\beta) = 0.021$ and $\sigma(\mu) = 0.039$. (b) Pattern with diffuse scattering observed in the specimen aged at 600 °C for 611 h. $\sigma(\beta) = 0.024$ and $\sigma(\mu) = 0.030$. (c) Pattern showing 6-fold like symmetry observed in the specimen aged at 700 °C for 130 h. $\sigma(\beta) = 0.19$ and $\sigma(\mu) = 0.094$.

In order to evaluate the quality of quasicrystals, two-dimensional positional data of reflections were measured in each electron diffraction pattern. We are interested in the shape of the quadrangle formed by the four reflections, $\boldsymbol{g}_1 = \bar{1}0320$, $\boldsymbol{g}_2 = \bar{1}0330$, $\boldsymbol{g}_3 = \bar{2}0330$ and $\boldsymbol{g}_4 = \bar{2}0320$. They form a square in the ideal case with the edge length $a^* = 0.1267$ Å^{-1}. Actually the vertical angles of the quadrangles were measured to be 90.0° with the standard error typically 0.2°. Then they are regarded as a rectangle in the diffraction patterns observed in this study. The rectangle has two edges, $a_x^* = |\boldsymbol{g}_1 - \boldsymbol{g}_4|$ and $a_y^* = |\boldsymbol{g}_3 - \boldsymbol{g}_4|$. Two parameters are introduced here: $\beta = a_y^* / a_x^*$ and $\mu = a_x^* / \overline{a^*}$ (or $\mu = a_y^* / \overline{a^*}$). Here $\overline{a^*}$ denotes the average of 48 edge lengths of 12 rectangles experimentally measured. The parameters β and μ express rectangularity and normalized edge length, respectively. The standard deviations of these parameters, $\sigma(\beta)$ and $\sigma(\mu)$, were calculated for 12 values of β and 48 values of μ experimentally measured, respectively. Both standard deviations are zero in the ideal case, if there is no experimental error. With respect to these standard deviations, it may be useful to describe two typical cases. One is the hexagonal approximant $Mn_{70}Cr_{10}Si_{20}$ with the lattice parameters a = 16.985 Å and c = 4.625 Å [4]. In this case, $\beta = 2/\sqrt{3}$ or $\sqrt{3}/2$, and the standard deviations are $\sigma(\beta) = 1/4\sqrt{3} \approx 0.144$, $\sigma(\mu) = 1 - \sqrt{3}/2 \approx 0.134$. The other is the tetragonal approximant, σ-phase, with the lattice parameters a = 8.806 Å and c = 4.624 Å. In this case, there are two kinds of squares with different edge lengths. The standard deviations are $\sigma(\beta) = 0$ and $\sigma(\mu) = (\sqrt{2}-1)^2 \approx 0.172$. In Figure 4, the two kinds of standard deviations are presented for the four specimens: A and B aged at 600 °C for 100 h and that aged at 600 °C for 611 h, and at 700 °C for 130 h. For this analysis, seven, five, nine, and twelve data sets were used respectively for each specimen. The specimen A aged at 600 °C for 100 h exhibits remarkably small deviations with respect to both β and μ. However, other specimens show larger deviations and wider distributions in $\sigma(\beta)$ and $\sigma(\mu)$. These results indicate tendency of structural fluctuation caused by longer aging time or higher aging temperature. Then the dodecagonal quasicrystal formed in this Mn-based alloy is possibly a metastable phase, while the growth behavior is well explained by Johnson-Mehl-Avrami Equation. The specimen B at 600 °C for 100 h shows larger standard deviations than the specimen A, and this difference clearly indicates the limitation of the present experiment.

The shift of reflections from the exact symmetric positions can be related to the tile rearrangement in the direct space that is called phason disorder [5]. The relationship between the standard deviations described here and the phason tensor is the subject to be studied.

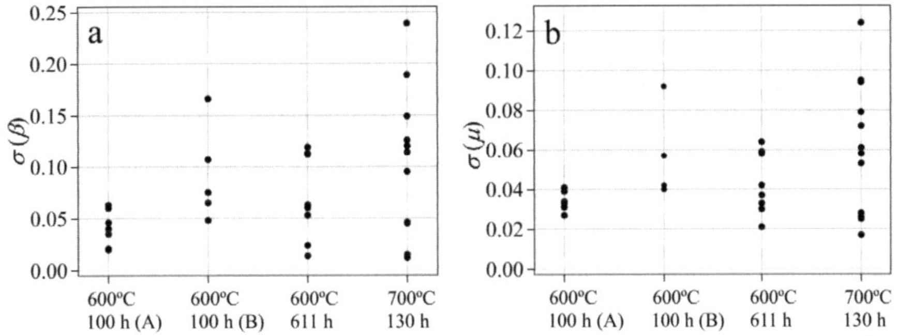

Figure 4. Deviations from the exact 12-fold symmetry showing high-quality of the specimen A aged at 600 °C for 100 h. (a) Standard deviation $\sigma(\beta)$. (b) Standard deviation $\sigma(\mu)$.

4. Conclusion

In the $Mn_{72.0}Cr_{5.5}Ni_{5.0}Si_{17.5}$ alloy, the dodecagonal quasicrystal is slowly grown from the matrix of β-Mn type phase by aging at 600 or 700 °C. This growth process is explained by Johnson-Mehl-Avrami Equation. However, the detailed analysis of electron diffraction patterns suggested the increase of the deviations from the ideal dodecagonal quasicrystal during the aging. In the limit of this experiment, the best condition to synthesize the dodecagonal quasicrystal is aging at 600 °C for 100 h. Even under this condition, the correlation length in the quasiperiodic direction is limited within 100 Å. Synthesis of the dodecagonal quasicrystal with high quality and uniformity is still beyond our control.

Acknowledgement
The authors thank Shuhei Iwami for his support in the specimen preparation.

References

[1] Iwami S and Ishimasa T 2015 *Phil. Mag. Lett.* **95** 229
[2] Ishimasa T, Iwami S, Sakaguchi N, Oota R and Mihalkovič M 2015 *Phil. Mag.* **95** 3745
[3] Yamamoto A 1996 *Acta Crystallogr.* **A52** 509
[4] Ishimasa T 2011 *Israel J. of Chem.* **51** 1216
[5] Leung P W, Henley C L and Chester G V 1989 *Phys. Rev.* **B39** 446

Preparation dependent surface structure of NiAl(100)

Sam Coates[1], Abdullah Al-Mahboob[1], Ronan McGrath[1] and Hem Raj Sharma[1]

[1]Department of Physics and Surface Science Research Centre, University of Liverpool, Liverpool L69 3BX, UK

E-mail: sgscoat2@liv.ac.uk, h.r.sharma@liv.ac.uk

Abstract. The dependence of surface structure formation on preparation conditions of NiAl(100) has been investigated by Scanning Tunnelling Microscopy (STM), Low Energy Electron Diffraction (LEED) and Density Functional Theory (DFT). STM and LEED have been used to study the surface after sputtering, low temperature annealing (T<500K) and high temperature annealing (500K<T<1000K). A (1×1) phase is observed both after sputtering and low T annealing, with STM images indicating the formation of row structures - the density of which appears dependent on annealing time. A $c(\sqrt{2} \times 3\sqrt{2})R45^{\circ}$ regime is detected upon higher T annealing - forming two, orthogonal, row-based domains. STM simulation produced using DFT explains the origin of one of the domains - a defect based structure with a dominant Ni density of states contribution.

1. Introduction

Nickel-Aluminium (NiAl) is a periodic CsCl-type intermetallic alloy which has been routinely investigated due to its advantageous physical properties such as high melting temperature, high strength, and good oxidation resistance. As such, NiAl alloys have been used as a base in superalloys for applications in, for example, corrosion resistant coatings and turbine blades [1]. Additionally, NiAl makes for an attractive and appropriate system for research on the general phenomena in ordered alloys due to its simplistic crystallographic structure and, by natural extension, the surface reconstructions produced from bulk termination [1]. The NiAl(100), (110) and (111) surface terminations can be regarded as prototypical systems for studying oxide growth and nanocluster formation in intermetallic compounds – an area of research with potential for applications in catalysis and electronic devices [2–6]. As such, a deep understanding of the surface structure of NiAl is imperative when considering the implications of such research.

The NiAl(100), (110) and (111) surface terminations, both clean and oxidised, have been investigated using a variety of techniques such as Low Energy Electron Diffraction (LEED) [4, 7, 9], Density Function Theory [10, 14] and Neutral Impact Collision Ion Scattering Spectroscopy (NICISS) [9]. Of specific interest here is the NiAl(100) termination - the (100) plane ideally produces alternate A–B–A–B layers of quadratically arranged pure Ni(A) or Al(B) atoms (a_0 = 2.89Å [8]). However, an energetically preferential reconstruction forms in an Al–terminated $c(\sqrt{2}\times 3\sqrt{2})R45^{\circ}$ regime when NiAl is annealed within a temperature range of 500 - 1000K [7, 8]. A (1×1) phase can be produced either by annealing at T < 500K (Al terminated) or at T > 1000K (Ni terminated) [7, 8]. Geometric optimisation of a defect rich NiAl(100) surface using Density Functional Theory (DFT) has revealed the origin of the reconstruction - concluding that there are two equally favourable defect-based atomic arrangements [10]. The first is a structure described as Ni antisites along every third row in the (011) direction, with corresponding vacancies in the Ni layer below - known as the 'double defect' surface from now onwards. The second is a 'single defect' regime solely exhibiting the Ni vacancies within the subsurface layer. The structures are shown in figure 3d) and e) respectively. The reconstruction is, energetically, the most stable surface structure [10].

Using STM, LEED, and DFT, both phases of the NiAl(100) surface have been studied. Presented here are new data from each stage of sample preparation in an effort to uncover how atomic surface structure is affected by such preparation techniques, and to explain the origin of both surface phases. NiAl has been studied as part of a larger programme looking at adsorption and thin film structure dependence on substrate structural complexity. This will include complex and quasicrystalline Al-based intermetallics, such as Al_9Co_2 and the 10-fold Al-Ni-Co.

2. Experimental Method

A NiAl(100) single crystal was polished to a mirror finish before solvent washing. A clean surface was prepared by Ar^+ sputtering for 20-30 minutes at pressures of $\sim 2.5\times10^{-5}$ mbar, followed by annealing at temperatures dependent on surface selection (T $<$ 500K for (1×1), T = 600 – 800K for reconstruction) at pressures below 5×10^{-10} mbar. The sputtering conditions were chosen to remove surface and subsurface ordering. Surface cleanliness and structural determination were verified using LEED and STM at room temperature. All bias voltages used in STM measurements are with respect to the sample. STM simulations were completed using CASTEP [11], employing the Perdew-Burke-Ernzerhof generalized gradient approximation functional [12].

3. Results and Discussion

3.1. Sputtered surface

To understand the NiAl(100) surface formation and the full effect of the preparation methods, the sputtered surface was investigated. The (1×1) diffraction pattern shown in figure 1a) (confirmed by comparison to Cu(111)) indicates long range order across the surface, despite the heavy sputtering treatment. Figure 1b) shows terrace formation after sputtering, with figure 1c) showing a magnified area of a terrace. Island features are observed with an RMS roughness of 0.14 nm. Line profile analysis (labelled 1 and 2) indicates that these islands are Al-Al or Ni-Ni, based on step heights of ~3Å ($\simeq a$).

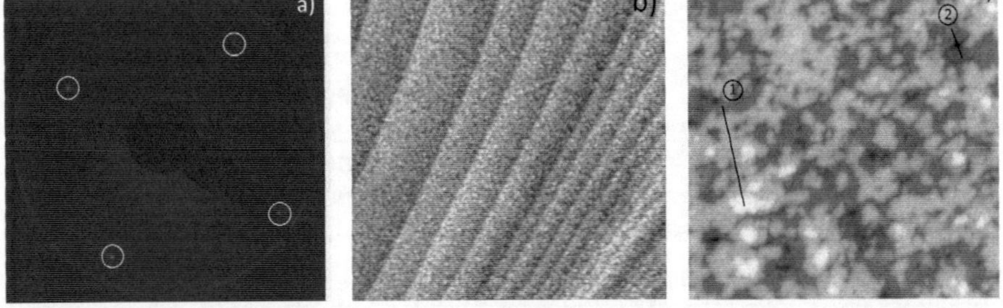

Figure 1: a) LEED pattern (E_e = 50 eV) from the sputtered NiAl(100) surface, producing a (1×1) diffraction pattern b) 500×500 nm^2 STM image (V_b = 2000 mV, I_t = 270 pA) exhibiting terraces c) 100×100 nm^2 STM image (V_b = 1750 mV) with labelled line profiles used to measure step heights of islands

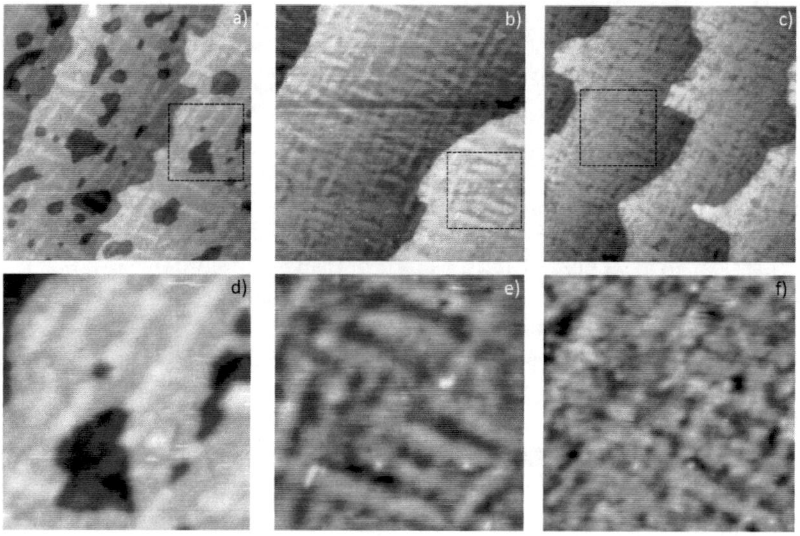

Figure 2: Series of STM images taken from different annealing preparation conditions. a) 100×100nm^2 STM image (V_b = -500mV, I_t = 270pA) annealing temperature T = 420K, b) 100×100nm^2 (V_b = 2000mV, I_t = 270pA), T = 470K, c) 100×100nm^2 (V_b = 150mV, I_t = 270pA), T = 480K, d-f) show 32×32nm^2 sections taken from highlighted regions in a-c) respectively

3.2. Low temperature annealed surface

Figure 2) shows a collection of STM scans from the (1×1) NiAl(100) surface (as verified by LEED, figure 3a)); all images are produced by annealing NiAl(100) below the reconstruction transition temperature, 500K [9], for increasingly longer amounts of time – 2a) 20 min, b) 40 min, c) 60 min. Terraces are formed in each procedure, although 2a) exhibits a high concentration of large defects in addition to roughened step edges. The non-square shape of these defects indicates either that the annealing temperature or length of annealing time was insufficient to produce a surface in equilibrium. In all 3 scans 'row' structures appear in two orthogonal directions and are highlighted by enlarged images 2d-f) – the density of these structures appears to increase with increasing annealing time. In fact, upon annealing for a long time (over 12 hours) at T<500K, the surface reconstructs into its $c(\sqrt{2} \times 3\sqrt{2})R45°$ phase - indicating that the appearance of these rows is a precursor to the formation of the reconstruction.

3.3. Reconstructed surface

Upon annealing at 500K<T<900K the NiAl(100) surface reconstructs into the aforementioned $c(\sqrt{2} \times 3\sqrt{2})R45°$ regime. Figures 3a) and b) show the LEED patterns from the (1x1) and reconstructed surface structure phases respectively. Included in figure 3c) is a LEED simulation of the reconstruction for clarity, where two domains oriented 90° to each other are clearly visible.

STM data from this surface also appears to exhibit two orthogonal domains, as shown in 3f) and highlighted by a dashed line. To the left of the line a defined row structure is observed with separation 6.1±0.2Å, whilst inter-atomic separation along these rows is difficult to resolve. The domain to the right of the line again exhibits row structure, yet separation between these rows appears to vary between ∼4.5-5.5Å. Additionally, ordering between adjacent rows appears to be arbitrary, as consecutive rows are shifted along their axis, causing a 'wavy' effect. The possible origin of the left-hand domain has been obtained by simulating the electron density of the double defect reconstruction as seen by an STM tip, using DFT; figure 3h). Here, the double defect structure has been replicated over 3 atomic layers – an Al termination with Ni double

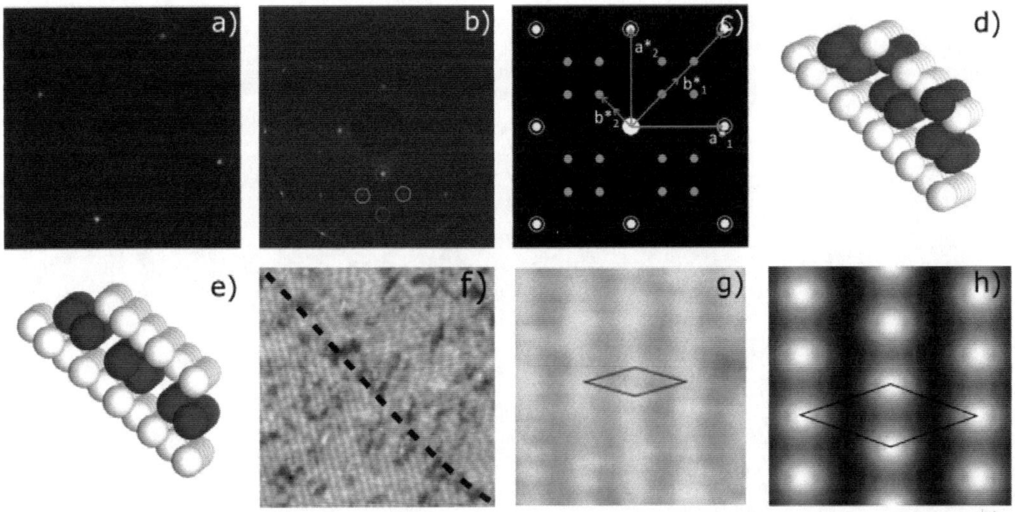

Figure 3: a) (1×1) LEED pattern (E_e = 50 eV) produced from NiAl(100) annealed at 470K, b) c($\sqrt{2} \times 3\sqrt{2}$)R45° LEED pattern ($E_e$ = 52 eV) produced from the (100) surface, annealed at 640K, c) simulated LEED pattern of the reconstruction, where red and blue dots correspond to two orthogonal domains [13], d) double defect structure e) single defect structure f) 15×15nm^2 STM image with two clearly observable domains, g) 3.0×2.5nm^2 region with proposed unit cell highlighted h) simulated STM from double defect structure, using parameters equal to those used in the physical STM. Proposed unit cell highlighted

defects within the top 2 layers, with a pure Al layer underneath. The virtual tip is placed ~3Å from the surface, with an applied bias of 175mV (probing unoccupied states) - equal to that used in the physical experiment. The STM simulation produced correlates with the STM data, and can be described as the Ni density of states (DOS) dominating at the surface, producing the defined row structure as described above. DOS calculations confirm a dominant Ni *d* band within NiAl [15].

4. Conclusion

NiAl(100) surface formation under various preparation conditions has been studied. The (1×1) surface ordering is maintained after heavy sputtering treatment, despite a morphologically rough surface – as evidenced by LEED. Annealing the sample at T < 500K produces a (1×1) surface, with orthogonal 'row' islands observed in STM – the density of which appears to increase with annealing time. Surface reconstruction to a c($\sqrt{2} \times 3\sqrt{2}$)R45° structure occurs after annealing at T > ~ 800K, which is in part attributed to a Ni based double defect structure in the subsurface, as shown by DFT calculations. Further studies will focus on the formation of islands on the (1×1) surface, including their part in formation of the reconstruction. In addition, the origin of the second domain in the reconstructed surface will be explored.

References

[1] Miracle D B, 1993 *Acta metall. mater.* **41** 649 – 84
[2] Qin H, Chen X, Li L, Sutter P W and Zhou G 2014 *PNAS.* E103 – 09
[3] Cai N, Qin H, Tong X, Zhou G 2013 *Surface Science* **618** 20 – 26
[4] Jaeger R M, et al. 1991 *Surface Science* **259** 235 – 52
[5] Luo M F, Chiang C I, Shiu H W, Sartale S D and Kuo C C 2005 *Nanotechnology* **17** 360 – 7
[6] Franchy R, Masuch J and Gassmann P 1996 *Applied Surface Science* **93** 317 – 27
[7] Mullins D R, Overbury S H 1987 *Surface Science* **199** 141 – 153
[8] Blum R P, Ahlbehrendt D, Niehus H 1996 *Surface Science* **366** 107 – 120
[9] Niehus H, Raunau W, Besocke K, Spitzl R and Comsa G 1990 *Surface Science* **225** L8 – L14
[10] Lerch D, Dössel K, Hammer L and Müller S 2009 *J. Phys.: Condens. Matter* **21** 134007 (6pp)
[11] Clark S J et al. 2005 *Zeitschrift fuer Kristallographie* **220** 567 – 70

[12] Perdew J P, Burke K and Ernzerhof M 1996 *Phys. Rev. Lett.* **77** 3856
[13] http://www.fhi-berlin.mpg.de/KHsoftware/LEEDpat/
[14] Müller S 2003 *J. Phys.: Condens. Matter* **15** 1429 – 1500
[15] Lui S C, Davenport J W, Plummer E W, Zehner D M, and Fernando G W 1990 *Phys. Rev. B* **42** 1582 – 1597

Medium energy ion scattering (MEIS) study from the five-fold surface of icosahedral Ag-In-Yb quasicrystal

H. R. Sharma[1], P. J. Nugent[1], T. C. Q. Noakes[2], J. Smerdon[1], J. Parle[1], A. P. Tsai[3,4], and R. McGrath[1]

[1]Surface Science Research Centre and Department of Physics, University of Liverpool, Liverpool, L69 3BX, UK
[2]STFC Daresbury Laboratory, Warrington, WA4 4AD, UK
[3]Institute of Multidisciplinary Research for Advanced Materials, Tohoku University, Sendai, 980-8577, Japan
[4]National Institute for Materials Science, 1-2-1 Sengen, Tsukuba, Ibaraki, 305-0047, Japan

E-mail: h.r.sharma@liv.ac.uk

Abstract. Medium energy ion scattering (MEIS) is employed to characterize the composition and structure of the five-fold surface of the icosahedral $Ag_{42}In_{42}Yb_{16}$ quasicrystal. The composition of the surface after sputtering is dominated by Ag and In, and when the surface is annealed at temperatures approaching 430°C, Yb is restored at the surface. The composition is that expected from the bulk structure if the surface is formed at bulk planes involving the centre of rhombic triacontahedral clusters, the building blocks of the system. Structural analysis of MEIS results are also consistent with a surface after annealing that is in close agreement with bulk truncation intersecting the cluster centre.

1. Introduction

The discovery of the binary icosahedral (*i*-)Cd-Yb quasicrystal[1] has led to the development of a series of isostructural ternary quasicrystalline and approximant systems that are suitable for surface studies under ultra-high vacuum (UHV). These systems include *i*-Ag-In-Yb and Ag-In-RE 1/1 approximants (RE: rare earth element Yb, Gd, Yb). The surface structure of these materials has been investigated by scanning tunneling microscopy (STM)[2-7]. The STM images were found to be either dependent on bias voltage[2] or atomic resolution was not obtained[5,6], making it difficult to analyse the images in terms of a bulk structure model. In this work, we employed medium energy ion scattering (MEIS) to characterise the structure and chemical composition of the five-fold surface of *i*-Ag-In-Yb.

The main building blocks of the *i*-CdYb system are rhombic triacontahedral (RTH) clusters which consist of five successive atomic shells (an outer-most RTH shell composed of 92 Cd atoms, a 30-atom Cd icosidodecahedron, a smaller 12-atom Yb icosahedron, a smaller 20- atom Cd dodecahedron, and an inner 4-atom Cd tetrahedron)[8]. The cluster atoms comprise 93.8% of the total atoms in the crystal with the remaining 'glue' atoms filling the space between the RTH clusters. Cd is replaced by an equal amount of Ag and In in *i*-Ag-In-Yb. However, the exact distribution of Ag and In in the RTH cluster is still unknown. STM studies show that all the three high symmetry surfaces (five-fold, two-fold and three-fold) of *i*-Ag-In-Yb truncate at the bulk planes intersecting the centre of the RTH cluster. However, the surface termination of the approximants is still under debate. The MEIS results for the icosahedral material are consistent with surface termination at cluster centres.

2. Experimental Details

The single grain sample of *i*-Ag-In-Yb (*i*-$Ag_{42}In_{42}Yb_{16}$) was grown using the Bridgman method[7]. The sample was cut perpendicular to the three-fold axis and then mechanically polished with successively finer grades of diamond suspension down to 0.25 μm. Subsequently, the surfaces were cleaned in an ultra-high vacuum (UHV) chamber with a base pressure of 2×10^{-10} mbar using repeated cycles of Ar^+ sputtering at 3 KeV for 30 mins and annealing at different temperatures up to 430°C for 2 hours each. MEIS data were taken after each annealing.

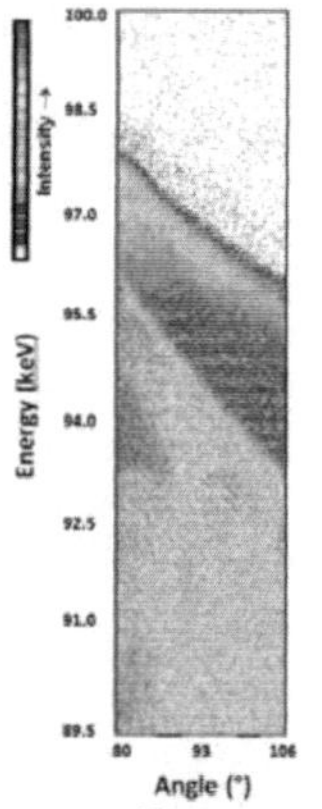

Figure 1. Two-dimensional MEIS spectra obtained from the clean five-fold *i*-Ag-In-Yb surface.

The MEIS experiments were carried out at Daresbury Laboratory[10] with a 100 keV He^+ ion beam incident on the surface. The angle and energy of the scattered ions were determined using a toroidal electrostatic energy analyser with position-sensitive detector and a data acquisition system which produced 2D spectra such as that shown in Fig. 1. The variation in scattered ion intensity with both scattered energy and angle is represented by a false colour map. By gating specific areas of the two-dimensional spectra and projecting onto either the energy or angle axis, one-dimensional spectra can be produced. Energy spectra typically provide information on the depth dependent composition, while angular spectra provide crystallographic information.[11]

3. Results and Discussion

Figure 1 shows a typical 2D MEIS spectrum from the five-fold surface of *i*-Ag-In-Yb after annealing. 1D energy and angle cuts from these 2D spectra are shown in Figures 2 and 3 and have been analysed to determine the composition and structure of the top-most layers and sub-surface region just below.

3.1. Surface composition

Figure 2(a) show 1D MEIS energy spectra taking from the five-fold *i*-Ag-In-Yb surface after sputtering and annealing at different temperatures. The characteristic features observed in the spectra are result of ions scattered from Ag, In and Yb. The peak at lower energy represents both Ag and In due to their similar mass. Only the small step at approximately 90.5 keV provides information on the relative contributions of In and Ag and then only for the top-most layers. As such, whilst the proportion of Yb can be determined with confidence, In and Ag have much larger relative uncertainty.

Figure 2(b) shows the normalised intensities of these elements with respect to their cross-sections (effectively the square of their atomic number) for the top-most layers (within around 4 Å of the surface). Figure 2(c) shows similar data for the sub-surface region (for these measurements, typically 10 to 15 Å below the top-most layers). The concentration of Yb after sputtering is about 15% in the sub-surface region and about 10% in the top-most layers, which is less than bulk composition, suggesting that Yb is selectively removed. The preferential sputtering of Yb and In was also observed by XPS from the same surface[12]. Both of these elements are heavier than Ag, however, their surface free energy is less. Therefore, the selective sputtering may be related to surface energy of the constituent elements rather than their mass. Elements with low surface free energy are likely to segregate to the surface and can then be sputtered by the incoming ions. The *i*-Al-Pd-Mn[13] and decagonal Al-Cu-Fe[14] quasicrystals show a depletion of Al upon sputtering. Al is the lightest element but also has the lowest surface free energy among the constituents in both systems. Another factor that may cause the selective sputtering is surface binding energy of the elements[14].

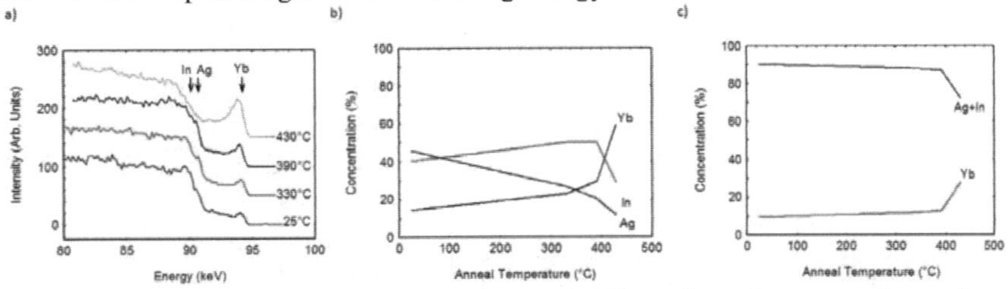

Figure 2. (a) 1D energy spectra from the five-fold *i*-Ag-In-Yb surface after sputtering and annealing at different temperatures. Peaks relating to Yb, Ag and In are highlighted. (b) Concentrations of Yb, Ag and In in the top-most layers versus annealing temperature extracted from (a) by fitting with the SIMNRA simulation program. (c) Concentrations of Yb and Ag + In combined for the sub-surface region

As the annealing temperature increases, the concentration of Yb increases and Ag/In decreases. At 430°C, the concentration of Yb is roughly 30% in the sub-surface region and about 60% in the top-most layers. These values are higher than the bulk concentration, i.e., also higher than the concentration determined by XPS[12]. For the top-most layers, MEIS is more surface sensitive than XPS, which averages over a slightly larger depth range. The composition of the top-most layers determined from the model structure is $Ag_{11.5\pm2.1}In_{45.7\pm8.7}Yb_{42.8\pm10.8}$. The model surface was selected at bulk planes which intersect the cluster centre and the concentration was determined considering atoms in a slab of 0.5 Å thickness are in a single plane. This thickness was chosen because the maximum projected atomic density in this region is comparable to that of the closed-packed Yb surface. The concentration of Yb would go down if a thicker slab is considered. The measured concentration of Yb in the top-most layers is above this, but the sub-surface region is slightly less than the value expected from the model. This might imply that the sample was annealed to a slightly higher temperature than desirable to obtain the ideal surface compositional profile. As previously stated, the close masses of Ag and In make them difficult to resolve in the MEIS spectra, precluding a similar analysis.

3.2. Structural Analysis

Angular spectra from the quasicrystalline sample after annealing are shown in Figure 3. The spectra exhibit significant noise as a result of the low flux of ions incident on the surface during the experiment. However, for the sub-surface data the three major blocking dips are still observed at 69°, 90° and 110.5°. An angular spectrum has been simulated using the VEGAS analysis code[13] and is included as Figure 3(c) for comparison. For the simulation, we used the atomic positions from the ideal model of the *i*-Cd-Yb quasicrystal[8] with the model containing an 8 Å slab. Due to the complexity of the model, the theoretical spectrum was limited to an angular resolution of 0.5°. The major blocking dips observed by experiments are reproduced in the calculation both in terms of position and amplitude, evidencing a close similarity between the theoretical and actual atomic positions.

The data from the top-most layers also show the same blocking dips, but with reduced amplitude. This may in part be because of greater illumination of these layers, which is typical for all materials, but enhanced for aperiodic materials where shadowing of the underlying atoms by surface atoms is less. Another possible contribution to the reduced amplitude is the presence of some Yb atoms that are not sitting in the predicted lattice sites, since this would be consistent with the excess Yb at the surface as evidenced by the compositional data. There is also a slight shift in the positions of the blocking dips towards lower scattering angle. This is evidence that some of the atoms in the top-most layers have relaxed inward towards the bulk. From simple geometric calculations the likely contraction can be deduced to be approximately 8% or around 0.1 Å. However, the lack of any additional features in the surface data suggests that no gross reconstruction of the surface. The MEIS data are consistent with an essentially bulk-like termination at the RTH cluster centres, but with a slight inward relaxation of the outermost atoms. This bulk-like termination is in agreement with STM observations[2].

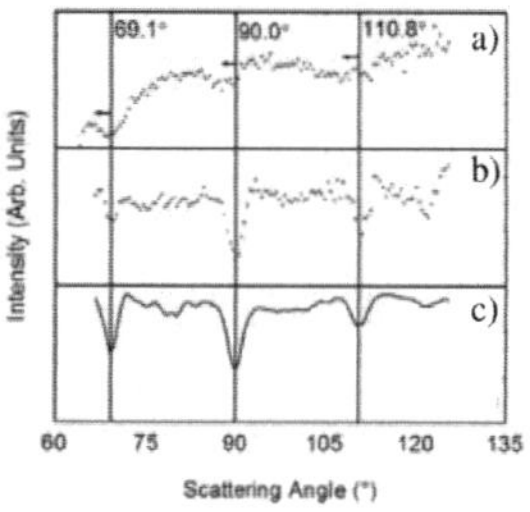

Figure 3. One-dimensional spectra taken along the angle axis, showing the intensity of scattered ions as a function of angle. (a) For the top-most layers (b) for the sub-surface region and (c) a simulated spectrum of the same material.

4. Conclusions

MEIS was used to study the atomic composition and structure of the five-fold surface of *i*-Ag-In-Yb. The surface after sputtering shows Yb depletion which is restored after annealing. This composition is expected for the surface formed at bulk planes intersecting the centre of RTH clusters, the building blocks of the system. Structural analysis is also consistent with bulk truncation at these planes, but with a slight inward relaxation of the top-most atoms.

Acknowledgements
This work is supported by EPSRC under Grant No. EP/ D071828/1.

References

[1] A P Tsai, J Q Guo, E Abe, H Takakura, and T J Sato 2000 *Nature* **408**, 537
[2] H R Sharma, M Shimoda, K Sagisaka, H Takakura, J A Smerdon, P J Nugent, R McGrath, D Fujita, S Ohhashi, and A P Tsai 2009 *Phys. Rev. B* **80**, 121401(R)
[3] C Cui, P J Nugent, M Shimoda, J Ledieu, V Fournée, A P Tsai, R McGrath, and H R Sharma 2012 *J. Phys.: Condens. Matter* **24**, 445011
[4] P J Nugent, J A Smerdon, R McGrath, M Shimoda, C Cui, A P Tsai and H R Sharma 2011 *Phil. Mag.* **91**, 2862
[5] C Cui, P J Nugent, M Shimoda, J Ledieu, V Fournée, A P Tsai, R McGrath, and H R Sharma 2014 *J. Phys.: Condens. Matter* **26** 015001
[6] C Cui, P J Nugent, M Shimoda, J Ledieu, V Fournée, A P Tsai, R McGrath and H R Sharma 2012 *J. Phys.: Condens. Matter* **24**, 445011
[7] S S Hars, H R Sharma, J A Smerdon, T P Yadav, A Al-Mahboob, J Ledieu, V Fournée, R Tamura, and R McGrath 2016 *Phys. Rev. B* **93**, 205428
[8] H Takakura, C P Gomez, A Yamamoto, M de Boissieu, and A P Tsai 2007 *Nature Materials* **6**, 58
[9] C Cui and A P Tsai 2010 *J. Cryst. Growth* **312**, 131
[10] P Bailey, T C Q Noakes, and D P Woodruff 1999 *Surf. Sci.* **426**, 358
[11] J.F. van der Veen 1985 *Surf. Sci. Rep.* **5**, 199
[12] H R Sharma, M Shimoda, S Ohhashi, and A P Tsai 2007 *Phil. Mag.* **87**, 2989
[13] C J Jenks, J W Burnett, D W Delaney, T A Lograsso, and P A Thiel 2000 *Appl. Surf. Sci.* **157**, 23
[14] J A Barrow, V Fournée, A R Ross, P A Thiel, M Shimoda, and A P Tsai 2003 *Surf. Sci.* **539**, 54
[15] J F Frenken, J F van der Veen and R M Tromp 1986 *Nucl. Inst. Methods Phys. Res. B* **17**, 334

Adsorption structure of Bi on the fivefold surface of i-Ag-In-Yb quasicrystal

Kazuki Nozawa[1] **and Yasushi Ishii**[2]

[1] Department of Physics and Astronomy, Kagoshima University, Korimoto, Kagoshima 890-0065, Japan

[2] Department of Physics, Chuo University, Kasuga, Tokyo 112-8551, Japan

E-mail: kazuki.nozawa@gmail.com

Abstract. The initial stage of the growth of a Bi overlayer on the fivefold surface of icosahedral Ag-In-Yb quasicrystal is studied using the first-principles density functional calculation. Comparing to the previous study about Pb adsorption on the same surface, it turns out that Bi is bound stronger than Pb on the surface, and small differences are found in the relative magnitude of the adsorption energy of some stable sites. Those may lead some differences in the sequence of the growth in the initial stage, it is concluded that the under- and first-layers observed in the growth of Pb are formed also in the Bi case.

1. Introduction

After the discovery of stable quasicrystals (QCs) [1], studies on the QC surfaces were started by several groups. Studies of clean surfaces revealed an important feature that, after proper sputter-annealing treatment, the surfaces recover the quasi-periodic atomic arrangement same as the bulk [2, 3, 4]. It stimulated studies that aim at fabricating single element QC using the existing QC surfaces as a template. Following the success by Franke *et al.* [5], quasiperiodic monolayers have been obtained in several systems so far [6, 7, 8, 9, 10].

In addition to these development, we recently succeeded in fabricating quasi-periodic multi-layered Pb film on the fivefold surface of the icosahedral (i-)Ag-In-Yb quasicrystal, and revealed that the Pb overlayers grow with imitating the atomic structure of the substrate QC [11].

In this paper, we study the initial stage of the single element QC growth using Bi as the adsorbate instead of Pb used in the previous study.

2. Methods and Calculations

In this paper, we calculate the adsorption energy of Bi adsorbed on the clean surface of the fivefold surface of i-Ag-In-Yb based on the density functional theory (DFT). The quasicrystalline substrate is isostructural to the binary i-Cd-Yb QC, where Cd is replaced by Ag and In. In order to handle the aperiodic surface within the DFT, the fivefold surface of the i-Ag-In-Yb was approximated by a disk-shaped cluster extracted from a QC model by Takakura *et al* [12]. The adsorption energy is defined as

$$E_{\text{ads}}(\boldsymbol{r}) = E_{\text{substrate+atom}}(\boldsymbol{r}) - \Big[E_{\text{substrate}}(\boldsymbol{r}) + E_{\text{atom}}(\boldsymbol{r})\Big], \tag{1}$$

where $\boldsymbol{r} = (x, y, z)$ is the position of Bi considered (x and y are the lateral positions, and z is the distance with the surface), and $E_{\text{substrate+atom}}$ is the total energy of the system that Bi adsorbed on the substrate. $E_{\text{substrate}}$ and E_{atom} are the total energies of the isolated substrate and Bi atom, respectively. The adsorption energy is calculated on a 0.5 Å × 0.5 Å grid, and interpolated using the cubic spline method. No structural relaxation has been carried out for both adsorbate and surface atoms. The atomic position of the adsorbate is changed manually on the grid points one-by-one. It should be noted that the disk-shaped cluster is not common for all calculations, but it is re-defined as the cluster center is always located just bellow the adsorbate. The size of the disk-shaped cluster is 4Å thickness and 30Å diameter. A simulation box of 40 Å × 40 Å × 18 Å is used for the supercell calculation. The calculation is carried out using the VASP program package [13]. The projecter-augumented-wave potentials [14, 15] are used to describe the interaction between the electron and ionic core. The 4f-electrons of Yb atom is kept frozen in the core. The generalized gradient approximation is adopted for the exchange-correlation term [16]. Non-spin polarized calculations are used. In order to estimate the errors in the calculation, we compared the adsorption energies at several sites calculated using the following systems.

(i) 6 Å thickness and 30 Å diameter cluster + 40 Å × 40 Å × 18 Å simulation box
(ii) 4 Å thickness and 40 Å diameter cluster + 50 Å × 50 Å × 18 Å simulation box
(iii) 4 Å thickness and 30 Å diameter cluster + 50 Å × 50 Å × 18 Å simulation box
(iv) 4 Å thickness and 30 Å diameter cluster + 40 Å × 40 Å × 26 Å simulation box

According to these tests, it turns out that the adsorption energy is almost converged for the lateral size of the disc-shaped cluster and the size of the simulation box. Relatively large differences in the adsorption energy are found when we change the cluster thickness (i). The difference is less than 0.1 eV for most sites, and about 0.2 eV at worst. The differences originated in the lateral size of the cluster and the size of the simulation box are one oder of magnitude smaller than these. The cubic spline interpolation cause a large error at some sites where the adsorbate-surface interaction is strongly repulsive, but such site is not crucial to find out the stable adsorption sites, and, therefore, it does not affect our conclusion.

Since QC does not have the translational periodicity, there are infinite number of candidates for the surface top layer. Thus, we use a dense and Yb-rich surface as the surface top layer according to the experimental findings [17]. The fractionally occupied sites are treated in the same way with our previous paper [11]. Figure 1(a) is the atomic structure of the surface top layer used in the calculation. The red and green spheres denote the In and Yb atoms, respectively. A structure that In-decagon surrounded by ten Yb-pentagons highlighted in blue is commonly found in this plane. We focus on the area around this structure in this paper.

3. Results and Discussions

Figures 2 (a), (b) and (c) show the adsorption energies of Bi at z =1.5Å, 1.0Å and 0.5Å, respectively. Since there is a local fivefold rotational symmetry around the centre of the area considered, we reduced the number of the sites to be calculated using the symmetry. Namely, we calculated the adsorption energies only for the sites in the first quadrant (x, $y > 0$). The energies of the rest sites are represented by those of the symmetry equivalent sites in the first quadrant.

One can find ten adsorption pathways (dark blue region) forming a decagon around the centre of figure 2(a) (just inside the Yb-decagon). These adsorption pathways are classified into two groups labeled A and B by the symmetry and local environment. The numbers following the labels in the figures stand for the distance between the Bi atom and the surface top layer (ten

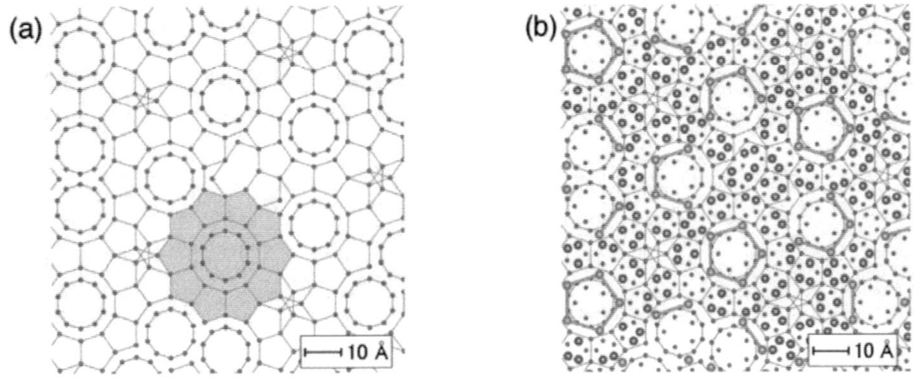

Figure 1. (a) The atomic structure of the topmost surface layer used in the calculation. (b) The adsorption structure of Pb overlaid on (a).

times z). The adsorption pathway labeled A extends vertically to the surface, and ends at around A10. On the other hand, the pathway B turns outward and takes the minimum value at around B05. Figure 3(a), which is the cross section of figure 2 along $y = 0$ line, shows it more clearly.

Similar structure of the adsorption energies can also be found in those of Pb adsorbing onto the same surface [11]. Figure 3(b) show the adsorption energies of Pb/i-Ag-In-Yb corresponding to figure 3(a) of Bi. In addition to the general tendency that Bi is bound to the surface stronger than Pb, the energy difference between A15 and B15 may make a difference in the sequence of the growth of the overlayer. The energy difference of the A15 and B15 sites of Pb is about 0.36 eV, and we concluded in previous report that the pentagonal *first-layer* composed of Pb atoms adsorbed on A10 sites are formed before formation of the *under-layer* comprised of Pb atoms adsorbed on B05 and other sites [11]. For Bi adsorbates, the energy difference between A15 and B15 is small (about 0.07 eV) to argue the stability of the adsorption pathway. Therefore, the sequence of the formation of the first- and under-layers of Bi can not be determined by these results.

However, the sequence of the formation of these layers may not be important for the final structure of the Bi overlayer, because, according to our previous calculations [18], the pre-adsorbed Pb on the A10 sites dose not prevent the next Pb atom from adsorbing onto the B05 sites. Thus, eventually, the first- and under-layers are expected to be formed like the Pb case shown in figure 1(b). Here the orange and purple spheres denote the adsorbate atoms occupied the A10 and B05 sites, respectively. Roughly estimated densities of A10 and B05 sites on the surface are about 0.6 $\times 10^{-3}$ Å^2 and 2.2 $\times 10^{-3}$ Å^2, respectively. The density of the A10 sites agrees very well to the experimental value for the Pb adsorbate (0.4 - 0.7 $\times 10^{-3}$ Å^2) [11].

Consequently, we conclude that there are no significant differences between the adsorption energies of Pb and Bi on the clean surface. Thus it is expected that Bi grow like Pb at least in its initial stage. An experiment carried out recently using scanning tunneling microscopy (STM) has observed different structures with the Pb case [19]. A possibility can be pointed out that the under- and first-layer of Bi have not been detected by STM like the under layer in the Pb case, and the observed layers may be formed on these undetected layers. Further theoretical investigations are required to check the hypothesis.

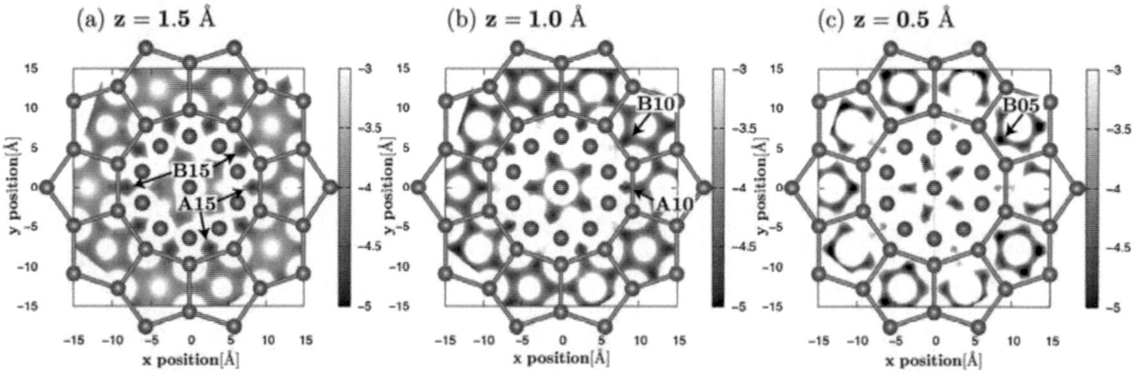

Figure 2. Calculated adsorption energies at (a) $z = 1.5$Å, (b) $z = 1.0$Å and $z = 0.5$Å. Characters A and B denote the prospective adsorption pathways (see text).

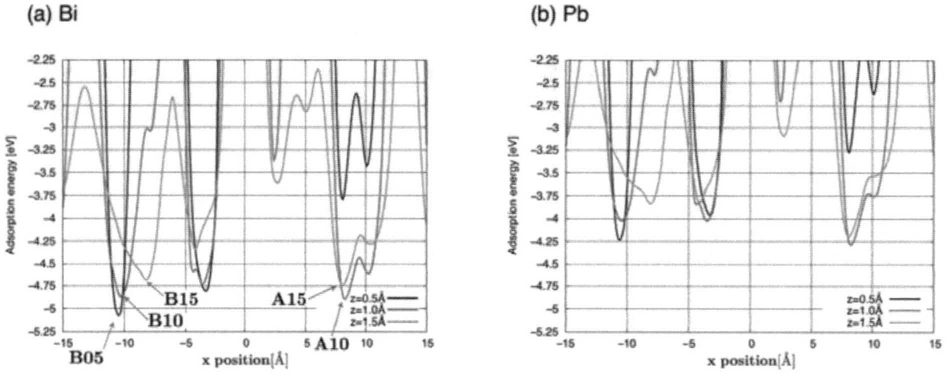

Figure 3. (a) Cross sections of figure 2 along the $y = 0$ line. (b) Same as (a) for Pb.

4. Concluding remarks

The stable adsorption structure on the clean surface of the fivefold surface of the i-Ag-In-Yb QC was studied using the first-principles density functional calculation. Calculated energy surfaces show that Bi atoms are bound stronger than Pb atoms, and imply slightly different adsorption behavior with Pb. However, those also indicate that the expected adsorption structure at the initial stage is very similar to the Pb case, where the pentagonal first-layer and the under-layer filling the interstitial region of the first-layer are formed.

Acknowledgments

The authors appreciate fruitful discussions with H. R. Sharma, M. Shimoda and A.P. Tsai. We are also grateful to H. Takakura for providing us with the atomic positions of *i*-Cd-Yb. Calculations are partly carried out using the supercomputer facilities at the National Institute for Materials Science, Tsukuba, and Yukawa Institute, Kyoto University.

References

[1] Tsai A P 2003 *Acc. Chem. Res.* **36** 31

[2] Ledieu J, McGrath R, Diehl R D, Lograsso T A, Delaney D W, Papadopolos Z and Kasner G 2001 *Surf. Sci* **492** L729-L734
[3] Sharma H R, Fournée V, Shimoda M, Ross A R, Lograsso T A, Tsai A P and Yamamoto A 2004 *Phys. Rev. Lett.* **93** 165502
[4] Sharma H R, Shimoda M, Sagisaka K., Takakura H, Smerdon J A, Nugent P J, McGrath R, Fujita D, Ohhashi S and Tsai A P 2009 *Phys. Rev. B* **80** 121401
[5] Franke K J, Sharma H R, Theis W, Gille P, Ebert Ph and Rieder K H 2002 *Phys. Rev. Lett.* **89** 156104
[6] Ledieu J, Leung L, Wearing L H, McGrath R, Lograsso T A, Wu D and Fournée V 2008 *Phys. Rev. B 77* 073409
[7] Deniozou Th, Ledieu J, Fournée V, Wu D M, Lograsso T A, Li H I and Diehl R D 2009 *Phys. Rev. B* **79** 245405
[8] Smerdon J A, Leung L, Parle J K, Jenks C J, McGrath R, Founée V and Ledieu J 2008 *Surf. Sci.* **602** 2496-2501
[9] Sharma H R, Shimoda M, Ross A R, Lograsso T A and Tsai A P 2005 *Phys. Rev. B* **72** 045428
[10] Shukla A K, Dhaka R S, D'Souza S W, Singh S, Wu D, Lograsso T A, Krajči M, Hafner J, Horn K and Barman S R 2009 *Phys. Rev. B* **79** 134206
[11] Sharma H R, Nozawa K, Smerdon J A, Nugent P J, McLeod I, Dhanak V R, Shimoda M, Ishii Y, Tsai A P and McGrath R 2013 *Nature Commun.* **4** 2715
[12] Takakura H, Gomez C P , Yamamoto A, de Boissieu M and Tsai A P 2007 Nat. Mater. 6 58
[13] Kresse G and Furthmüller J 1996 *Phys. Rev. B* **54** 11169
[14] Blöchl P E 1994 *Phys. Rev. B* **50** 17953
[15] Kresse G K and Joubert J 1999 *Phys. Rev. B* **59** 1758
[16] Perdew J P, Burke K and Ernzerhof M 1996 *Phys. Rev. Lett.* **77** 3865
[17] Sharma H R, Shimoda M, Sagisaka K, Takakura H, Smerdon J A, Nugent P J, McGrath R, Fujita D, Ohhashi S and Tsai A P 2009 *Phys. Rev. B* **80** 121401(R)
[18] Nozawa K and Ishii Y 2011 *Philos. Mag.* **91** 2913
[19] Sharma H R *et al.* to be published.

IOP Conf. Series: Journal of Physics: Conf. Series **809** (2017) 012019 doi:10.1088/1742-6596/809/1/012019

Phason dynamics in decagonal quasicrystals

Hans-Rainer Trebin and Hansjörg Lipp

Institute for Theoretical and Applied Physics, University of Stuttgart, Stuttgart, Germany

E-mail: trebin@itap.uni-stuttgart.de

Abstract. In a spectacular experiment of the year 2000 Edagawa et al. observed white spots in HRTEM images of d-Al-Cu-Co, which formed the vertices of a tiling. On a time scale of seconds to minutes the spots changed positions corresponding to phason flips. To illuminate the origin of the spots, their jump mechanism and frequencies we employ a structure model of Zeger and Trebin of 1996. It consists of quasiperiodic double layers stacked periodically by 4.18 Å. Each double layer contains rings of ten atoms whose centers form the vertices of a Tübingen triangle tiling. Correlated jumps of two atoms in each sublayer cause a simpleton phason flip of the entire ten ring. The stacking of the double layers leads to columns of the ten rings which might be interrupted due to the flips. Columns which contain a critical minimum number of complete ten rings are considered visible as white spots. Simultaneous appearences of vertices and their flipped positions give rise to the pentagonal Penrose pattern as observed by Edagawa et al. and also in other experiments. Postulating microscopic flip rates for the correlated jumps of pairs of atoms in the sublayers we calculate the probabilities for the numbers of complete ten rings in a column and confirm the mesoscopic time scales for the jumps of the white spots.

1. Introduction

Phasons are elementary excitations peculiar to quasicrystals. The notion comprises alternate positions of vertices and their flips in quasicrystalline structures. If these vertices are occupied by atoms, they are separated by a few angstrøms, and the time between the flips is of the order of picoseconds. The first direct observation of phason flips in a quasicrystal was reported by Edagawa et al. in a widely noticed article of the year 2000 [1]. The authors detected white spots in HRTEM images of decagonal $Al_{65}Cu_{20}Co_{15}$, which formed the vertices of a pentagonal Penrose tiling and which performed phason flips. The spots themselves, however, had diameters and separations in the order of nanometers. The time between their flips ranged from seconds to minutes. In the present contribution we attempt to explain the spots and their jumps by taking recourse to a structural model for decagonal Al-Cu-Co established by Zeger and Trebin in 1996 [2]. It contains columns of rings with ten atoms whose projections form the vertices of a Tübingen triangle tiling. We interpret these columns as the positions of the white spots. By collective jumps of atoms a column or part thereof can perform a simpleton flip of the triangle tiling. If the part which has flipped is large enough, both the original white spot and the flipped position are visible. In this way the original Tübingen triangle tiling turns over into the pentagonal Penrose tiling, which Edagawa et al. [1] and others (for example Ref. [3]) claim to observe. By postulating microscopic flip rates for the jumps of individual atoms we calculate the probabilities for jumps of the white spots and confirm the mesoscopic time scales.

2. Model for the structure of d-Al-Cu-Co

2.1. Model of Zeger and Trebin

The structural model of Zeger and Trebin [2] which we apply here is a slight modification of a model, which Burkov proposed in 1993 [4] based on experimental data of Steurer of 1990 [5]. It consists of quasiperiodic double layers stacked periodically by 4.18 Å. The projection of each double layer is a complex decoration of the two golden triangles comprising the decagonal Tübingen triangle tiling (Fig. 1). There are 34 atoms in the large and 21 in the small triangle. The composition is $Al_{62}Cu_{19}Co_{19}$. Each vertex of the tiling is the center of a ring of ten atoms. These rings form a column along the stacking axis. We join the assumption in the literature that these columns are to be identified with the white spots. Zeger and Trebin [2] documented a self-similarity both for the structure and the dynamics of the model. The structure is equivalent to a three times deflated triangle tiling containing at most two atoms in a triangle. In the highest deflation step one finds closely neighbored oriented rhombs in each layer, consisting of two large and two small golden triangles (Fig. 1). There is an aluminum atom on the central vertex and one within one of the large triangles. Through correlated jumps of these two atoms, each by 0.87 Å, the inner vertex performs a simpleton flip and the rhomb reverses its orientation. If this flip is occurring in the rhomb of the lower and the upper layer simultaneously, then an entire ten-ring of atoms is jumping and performing a simpleton flip in the initial, undeflated tiling (Fig. 1).

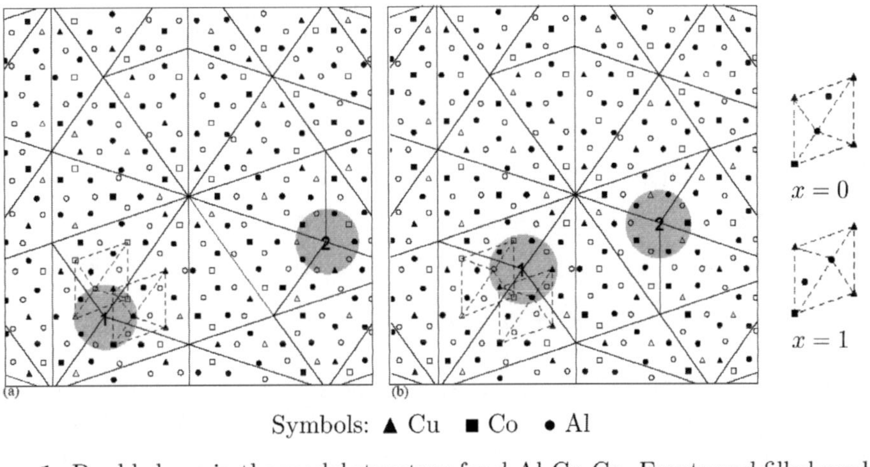

Figure 1. Double layer in the model structure for d-Al-Cu-Co. Empty and filled symbols denote atoms in the lower and upper layer, respectively. Each vertex is surrounded by a ring of ten atoms. The aluminum atoms of a ring belong to rhombs (dashed) of the deflated tilings (left figure). A rhomb can change from its original state ($x = 0$) to a reflected one ($x = 1$) by a simpleton flip. While doing so the Al atom inside the large triangle moves by 0.87 Å to the new vertex position; the previous vertex atom leaves by the same distance for the interior of the freshly formed large triangle. If both rhombs perform this flip (from state 00 to 11) an entire ten-ring is jumping in the original, undeflated tiling (right figure) [2].

2.2. Visibility criteria

Let us now consider a number N of (single) layers as in Fig. 2 and attach to each layer the state $x = 0$ or $x = 1$ of its small rhomb. Whenever we find successive states 00 or 11, a complete

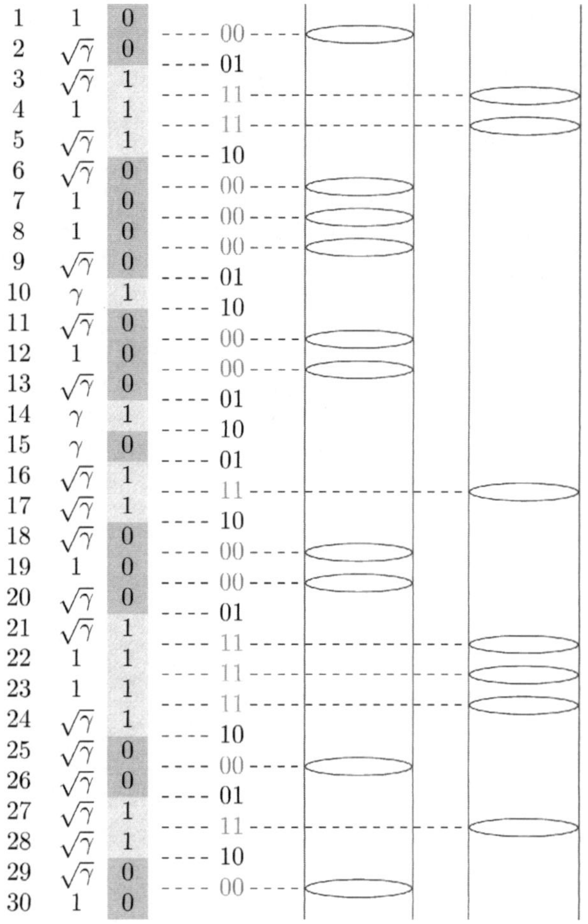

Figure 2. First column: numbering of the single layers. Second column: relative flipping probability of a single small rhomb. Third column: state of the small rhomb and cluster structure. Forth column and adjacent rings: state of the double layer, forming a ten-ring to the left if of type 00, to the right if of type 11.

ten-ring is found in the original or flipped state, respectively. We denote their numbers by n_{00} and n_{11} and introduce a visibility parameter α between 0 and 1. A white spot is assumed visible, whenever $n_{00} > \alpha N$ or $n_{11} > \alpha N$. When α is smaller than $\frac{1}{2}$ it might occur that both unflipped and flipped positions are visible. In the insert of Fig. 3 we display the original HRTEM image of Edagawa et al. (Fig. 2 of Ref. [1]) where we have imposed a part of the model of Zeger and Trebin [2] (black). The white lines denote the rhomb, pentagon and hexagon by which Edagawa et al. depicted a section of the pentagonal Penrose tiling before and after the flip (Fig. 3 of Ref. [1]). To the right of this insert the corresponding section of a Tübingen triangle tiling is drawn. The blue vertices are positions of the basic tiling, the green ones positions of simpleton flips, comprising a first deflation stage. The colored lines connect vertices corresponding to those of the HRTEM image. Further flips can proliferate the vertices and give rise to the pentagonal Penrose tiling.

3. Visibility probabilities and dynamics

For analyzing visibility probabilities of the white spots and from there the mesoscopic flipping times, the notion of clusters is very useful. A cluster is a consecutive series of identical states of the small rhombs in the single layers, see colored sections in the third column of Fig. 2.

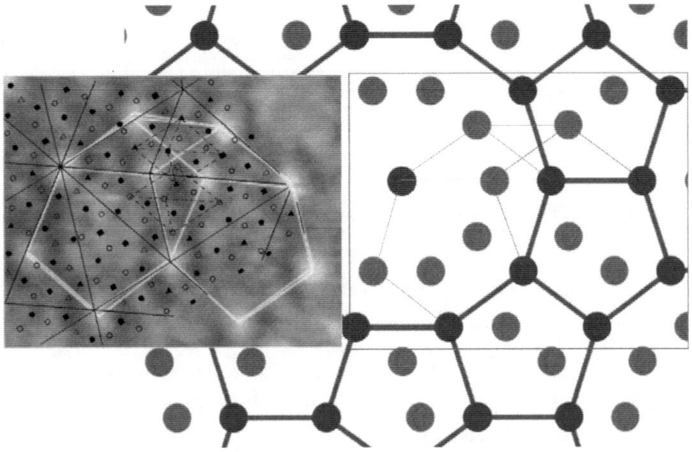

Figure 3. Left: HRTEM image of Edagawa et al. [1] with imposed tiles of the pentagonal Penrose pattern before and after the flip (white) and model of Zeger and Trebin [2] (black). Right: Corresponding section of a Tübingen triangle tiling (blue) and simpleton flipped positions (green).

If the number S_c of clusters is small, then the number of complete ten-rings is large, which becomes clear by the very simple relation $n_v := n_{00}+n_{11} = N-S_c$. To determine the probability for S_c clusters to occur we have associated relative jump rates to the small rhomb in a single layer. The highest rate $\gamma > 1$ is assigned, if the adjacent layers are in the same state, but different from the considered layer. If the rhombs in a layer and its neighbors are pointing to the same direction, the lowest jump rate $\gamma = 1$ is postulated. An intermediate rate $\sqrt{\gamma}$ is assumed, if the neighbors differ in their state. Thus we favor the formation of clusters.

By enumerating all flip processes which change the number of clusters in combination with kinetic Monte Carlo simulations we found that the numbers of clusters S_c are binominally distributed, as consequently are also n_v, n_{00} and n_{11}. The probabilities were calculated for one spot to appear, for two spots and for none. Finally the mesoscopic flip rates were derived. There are ranges of the jump rate γ and the visibility parameter α where these probabilities and the mesoscopic flip rates agree with the data of the video clips of Edagawa et al. [1]. Details are found in the thesis of Hansjörg Lipp [6].

References

[1] Edagawa K, Suzuki K and Takeuchi S 2000 *Phys. Rev. Lett.* **85** 1674
[2] Zeger G and Trebin H R 1996 *Phys. Rev.* B **54** R720
[3] Nagao K, Inuzuka T, Nishimoto K and Edagawa K 2015 *Phys. Rev. Lett.* **115** 075501
[4] Burkov S E 1993 *Phys. Rev.* B **47** 12325
[5] Steurer W 1990 *Acta Crystalogr.* B **46** 703
[6] Lipp H 2014 *Phasonendynamik in dekagonalen Quasikristallen* Ph.D. thesis University of Stuttgart

Sub-diffusive electronic states in octagonal tiling

G. Trambly de Laissardière†, C. Oguey† and D. Mayou*‡

† Laboratoire de Physique théorique et Modélisation, CNRS and Université de Cergy-Pontoise, 95302 Cergy-Pontoise, France
* Univ. Grenoble Alpes, Inst NEEL, F-38042 Grenoble, France
‡ CNRS, Inst NEEL, F-38042 Grenoble, France

E-mail: guy.trambly@u-cergy.fr

Abstract. We study the quantum diffusion of charge carriers in octagonal tilings. Our numerical results show a power law decay of the wave-packet spreading, $L(t) \propto t^{\beta}$, characteristic of critical states in quasicrystals at large time t. For many energies states are sub-diffusive, i.e. $\beta < 0.5$, and thus conductivity increases when the amount of defects (static defects and/or temperature) increases.

Experimental investigations have indicated that the conduction properties of many stable quasicrystals (AlCuFe, AlPdMn, AlPdRe, ...) are unusual and differ strongly from those of simple inter-metallic alloys [1-3]. In particular their conductivity increases with static defects density and when temperature increases. It appears also that the medium range order and the chemical order –over one or a few nanometers– have a decisive influence [4-9]. There is now strong evidence that these non standard properties result from a new type of break-down of the semi-classical Bloch-Boltzmann theory of conduction [10-14]. On the other hand, the specific role of long range quasiperiodic order in electronic properties is still an open question in spite of a large number of studies (Refs. [15-37] and Refs. therein). Many studies support the existence of critical states, which are neither extended nor localised, but are characterised by a power law decay of the wave-function envelope at large distances. In the presence of critical states, the diffusion of charge carrier at sufficiently large time t follows a power law and then the spatial extension L of wave-packets should be, [34-37]

$$L(E,t) \propto t^{\beta(E)} \text{ at large } t, \tag{1}$$

where β, $0 \leq \beta \leq 1$, is an exponent depending on energy E and on the Hamiltonian model. Note that in usual metallic crystals without static defects, $\beta = 1$ and the propagation is ballistic. In strongly disordered systems, $\beta = 0.5$ for a large time range and propagation is diffusive. For a localised state one has $\beta = 0$. When disorder is introduced in the perfect approximant or perfect quasicrystal in the form of static defects (elastic scatterers) and/or inelastic scattering (temperature, magnetic field...), the defects induce scattering and we expect that there is an associated time τ above which the propagation of the wave-packet is diffusive. The diffusivity D of charge carrier at energy E can be estimated by, $D(E,\tau) \simeq L(E,t=\tau)^2/\tau \propto \tau^{2\beta(E)-1}$, and the conductivity σ at zero frequency is given by the Einstein formula:

$$\sigma(E_F,\tau) = \mathrm{e}^2 n(E_F) D(E_F,\tau) \propto \tau^{2\beta(E_F)-1}, \tag{2}$$

where n is the density of states and E_F the Fermi energy. The case $0.5 < \beta < 1$, called super-diffusive regime, leads to transport properties similar to metal, since the conductivity decreases when disorder increases –i.e. when τ decreases–. Conversely, for $0 < \beta < 0.5$, the regime is sub-diffusive and the conductivity increases when disorder increases like in real quasicrystals. Many authors consider [18-20,29,34-37] that critical states could lead to $\beta < 0.5$ but it has not yet been shown in 2D or 3D quasiperiodic structures (except for some very specific energies).

Model Hamiltonian.– The octagonal, or Ammann-Beenker, tiling [38, 39] is a quasiperiodic tiling analogous to the notorious Penrose tiling. This tiling has been often used to understand the influence of quasiperiodicity on electronic transport [18-25]. A sequence of periodic approximants $X_0, X_1, \ldots, X_k, \ldots$ can be generated [40]. In approximants of order $k \geq 1$, the 6 local configurations around vertexes are the same as in the octagonal quasiperiodic tiling. They have, respectively, coordination number $\eta = 3, 4, 5, 6, 7$ and 8. We consider the simple Hamiltonian,

$$\hat{H} = \sum_i \epsilon_i c_i^* c_i + \sum_{\langle i;j \rangle} \gamma c_i^* c_j + h.c., \tag{3}$$

where i indexes s orbitals located on vertexes, and γ is the strength of the hopping between orbitals. $\langle i,j \rangle$ are the nearest-neighbours at tile edge distance a. To simulate schematically a possible effect of the presence of different chemical elements, the one-site energy ϵ_i is proportional to the coordinance η_i of the site i: $\epsilon_i = \eta_i \gamma$. To obtain realistic time values, we use $\gamma = 1\,\mathrm{eV}$ which is the order of magnitude of the hopping parameter in real inter-metallic compounds. The total density of states (total DOS) of X_7 approximant is shown figure 1(a).

Quantum diffusion.– In the framework of Kubo-Greenwood approach for calculation of the conductivity, we use the polynomial expansion method developed by Mayou, Khanna, Roche and Triozon [41, 42, 31, 27, 28] to compute the mean square spreading of the wave-packet at time t and energy E: $L^2(E,t) = \langle (\hat{X}(t) - \hat{X}(0))^2 \rangle_E$, where $\hat{X}$ is the position operator in the x-direction. The diffusion coefficient $\mathcal{D}(E,t) = L(E,t)^2/t$ is shown in figure 1(b) for X_7 approximant at some energies. The ballistic regime due to the periodicity of the approximant is reached at very large t, when $L(t) > L_k$ where L_k is the approximant cell size; then $L(t) = V_B t$, where V_B is the Boltzmann velocity, *i.e.* the intra-band velocity, $V_B = \langle \partial E_n(k)/\partial k_x \rangle_E / \hbar$, where $E_n(k)$ is the band dispersion relation [12]. For X_7 in time range shown figure 1, this Boltzmann term is negligible and, for all purposes of this discussion, the X_7 approximant is equivalent to the quasiperiodic system. Depending on the t values, three different regimes are observed at each energy:

- At very small time, typically when $L(t) < a$, the mean spreading grows linearly with t, $L(t) = V_0 t$ (ballistic behaviour), where $V_0 > V_B$ [10, 12].
- For times, corresponding to $L(t) \gtrsim$ a few a, the propagation seems to become diffusive as the diffusion coefficient is almost constant, $\mathcal{D}(t) \simeq \mathcal{D}_{dif}$. Therefore L_1,

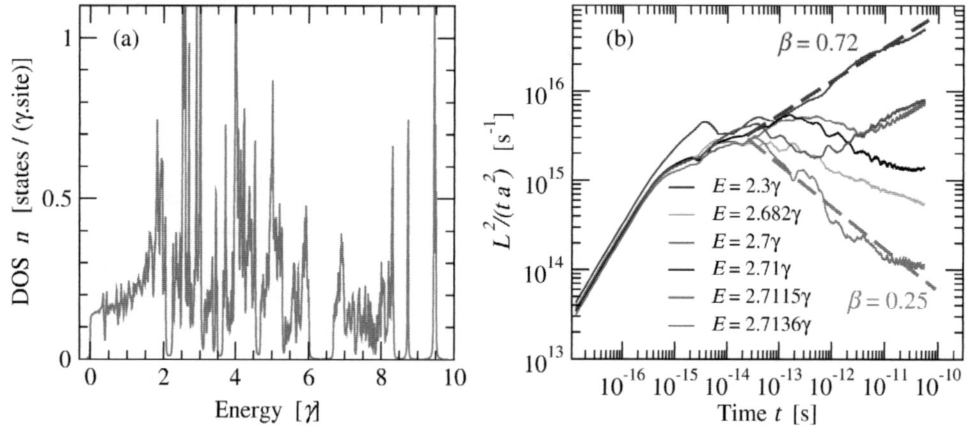

Figure 1. Density of states and quantum transport in perfect X_7 octagonal approximants (number of sites per cell $N = 275807$, unit cell $478a \times 478a$, with a the tile edge length): (a) Total DOS n (the recursion method induces a convolution of $n(E)$ by a Lorentzian of width 5 meV), (b) diffusive coefficient $\mathcal{D}(t) = L(t)^2/t$ for various E.

defined by $L_1 = \mathcal{D}_{dif}/V_0 \simeq$ a few a, is a kind of effective elastic scattering length but it is not due to static scattering events because we consider perfect tilings. The corresponding effective elastic scattering time is $t_1 = L_1/V_0$. Roughly speaking, it seems that when $L(t) \gtrsim L_1$, *i.e.* $t \gtrsim t_1$, the wave-packet feels a random tiling.

- An other distance L_2 (respectively an other time t_2, $L(t_2) = L_2$) appears. For $L(t) > L_2 \simeq$ a few $10a$, a new regime appears and $\mathcal{D}(t)$ follows a power law. It is thus characteristic of the medium and long range quasiperiodic order. Figure 1(b) shows that the β value can switch from a sub-diffusive regime ($\beta < 0.5$) to a super-diffusive regime ($\beta > 0.5$) over a small variation of energy. The t_2 values, $t_2 \simeq 10^{-13}$–10^{-14} s, have the order of magnitude of the scattering time above which measurements show unusual transport properties in quasicrystals [1].

Both distances $L_1 \simeq$ a few a, $L_2 \simeq$ a few $10a$, and the exponent β at time $t > t_2$, depend a lot on the energy value E. $L_1 < L_2$, but at some energy it even seems that $L_1 \simeq L_2$. Further analysis are necessary to understand the energy dependence.

To summarise, we have presented quantum diffusion in a large approximant of the octagonal tiling. The charge carrier propagation is determined by the wave-packet spreading in the quasiperiodic lattice. From numerical calculation, two length scales seem to characterise this quasiperiodic spreading. L_1, typically $L_1 =$ a few a, above which the propagation is almost diffusive in spite of the absence of static defects. L_2, typically $L_2 =$ a few $10a$, above which specific quasiperiodic symmetries lead to a power law dependence of the root mean square spreading, $L(t) \propto t^{\beta}$. For some energies states are super-diffusive or diffusive, *i.e.* $\beta \geq 0.5$, whereas for other energies, a sub-diffusive regime, *i.e.* $\beta < 0.5$, sets in as expected for critical states characteristic of quasiperiodicity. This sub-diffusive regime is the generalisation to quasicrystal of the non-Boltzmann propagation found in realistic approximants of i-AlMnSi and i-AlCuFe [10, 11], in the complex inter-metallic alloys λ-AlMn [12], and in small approximants of octagonal and Penrose tilings [13, 14].

The computations were performed at the Centre de Calcul (CDC), Université de Cergy-Pontoise. We thank Y. Costes for computing assistance.

References

[1] Berger C 1994 *Lectures on Quasicrystals*, ed F Hippert and D Gratias (Les Ulis, Les Editions de Physique) p 463
[2] Grenet T 2000 *Quasicrystals: Current Topics*, ed E Belin-Ferré, C Berger, M Quiquandon and A Sadoc (Singapore 2000, World Scientific) p 455
[3] Delahaye J and Berger C 2015 *Eur. Phys. J. B* **88** 102
[4] Fujiwara T, Yamamoto S and Trambly de Laissardière G 1993 *Phys. Rev. Lett.* **71** 4166
[5] Krajčí M et al. 1995 *Phys. Rev.* B **51** 17355
[6] Fujiwara T, Mitsui T and Yamamoto S 1996 *Phys. Rev. B* **53** R2910
[7] Roche S and Fujiwara T 1998 *Phys. Rev. B* **58** 11338
[8] Trambly de Laissardière G and Mayou D1997 *Phys. Rev. B* **55** 2890
[9] Trambly de Laissardière G, Nguyen Manh D and Mayou D 2005 *Prog. Mater. Sci.* **50** 679
[10] Trambly de Laissardière G, Julien J P and Mayou D 2006 *Phys. Rev. Lett.* **97** 026601
[11] Trambly de Laissardière G, Julien J P and Mayou D 2008 *Phil. Mag.* **88** 2131
[12] Trambly de Laissardière G and Mayou D 2014 *C. R. Physique* **15** 70
[13] Trambly de Laissardière G, Oguey C and Mayou D 2011 *Phil. Mag.* **91** 2778
[14] Trambly de Laissardière G, Szallas A and Mayou D 2014 Acta Phys. Pol. **126** 617
[15] Kohmoto M and Sutherland B 1986 *Phys. Rev. Lett.* **56** 2740
[16] Arai M, Tokihiro T, Fujiwara T and Kohmoto M 1988 *Phys. Rev. B* **38** 1621
[17] Fujiwara T, Kohmoto M and Tokihiro T 1989 *Phys. Rev. B* **40** 7413
[18] Sire C and Bellissard J 1990 *Europhys. Lett.* **11** 439
[19] Passaro B, Sire C and Benza V G 1992 *Phys. Rev. B* **46** 13751
[20] Sire C 1994 *Lectures on Quasicrystals*, ed F Hippert and D Gratias (Les Ulis, Les Editions de Physique) p 505
[21] Jagannathan A 1994 *J. Phys. I France* **4** 133
[22] Zhong J X and Mosseri R 1994 *J. Phys. I France* **4** 1513
[23] Zijlstra E S 2002 *Phys. Rev. B* **66** 214202
[24] Zijlstra E S 2004 *J. Non-Cryst. Solids* **334** 126
[25] Jagannathan A and Piéchon P 2007 *Phil. Mag.* **87** 2389
[26] Rieth T and Schreiber M 1995 *Phys. Rev. B* **51** 15827
[27] Roche S and Mayou D 1997 *Phys. Rev. Lett.* **79** 2518
[28] Roche S and Mayou D 1999 *Phys. Rev. B* **60** 322
[29] Mayou D 2000 *Phys. Rev. Lett.* **85** 1290
[30] Yamamoto S and Fujiwara T 1995 *Phys. Rev. B* **51** 8841
[31] Triozon F, Vidal J, Mosseri R and Mayou D 2002 *Phys. Rev. B* **65** R220202
[32] Jagannathan A, Szallas A, Wessel S and Duneau D 2007 *Phys. Rev. B* **75** 212407
[33] Szallas A and Jagannathan A 2008 *Phys. Rev. B* **77** 104427
[34] Guarneri I 1993 *Europhys. Lett.* **21** 729
[35] Guarneri I and Mantica G 1994 *Phys. Rev. Lett.* **73** 3379
[36] Piéchon F 1996 *Phys. Rev. Lett.* **76** 4372
[37] Schulz-Baldes H and Bellissard J 1998 *Rev. Math. Phys.* **10** 1
[38] Ammann R, Grunbaum B and Shephard G C 1992 *Discrete Comput. Geom.* **8** 1
[39] Socolar J E S 1989 *Phys. Rev. B* **39** 10519
[40] Duneau M, Mosseri R and Oguey C 1989 *J. Phys. A Math. Gen.* **22** 4549
[41] Mayou D 1988 *Europhysics Lett.* **6** 549
[42] Mayou D and Khanna S N 1995 *J Phys I France* **5** 1199

IOP Conf. Series: Journal of Physics: Conf. Series **809** (2017) 012021 doi:10.1088/1742-6596/809/1/012021

Screening of a magnetic impurity in a quasicrystal

Anuradha Jagannathan

Laboratoire de Physique des Solides, Université Paris-Sud, 91400 Orsay, France

E-mail: jagannathan@lps.u-psud.fr

Abstract. The problem of screening of local moments in a metallic environment is revisited, in the context of quasiperiodic solids. The motivation for this study comes from experimental studies showing that low temperature thermodynamic properties are strongly anomalous in the quasicrystalline phase, but less or not anomalous in the related approximant phase. A model for a magnetic impurity in a quasiperiodic environment is introduced and solved numerically. Arguments are given for the existence of a singular distribution of Kondo temperatures in the quasicrystal, thereby providing a possible explanation for the experimental observations.

1. Introduction

The discovery of a heavy fermion quasicrystal, YbAuAl, with anomalous thermodynamic properties at low temperature [1, 2] raised fundamental questions regarding the electronic states in metallic quasicrystals, and of the role played by electron-electron interactions in these materials. Already at the single particle level, numerous studies in a variety of models [3] have shown that quasiperiodicity has strong consequences for the nature of electronic states. Their spatial characteristics are very different from those of the usual Bloch states of a periodic system, bearing, instead, closer resemblances to states at the critical point of the metal-insulator transition in disordered metals. The local and global density of states have strong fluctuations, with pseudogaps of varying amplitudes. These fluctuations in energy and in position can be expected to have experimentally detectable consequences. One consequence concerns transport properties, and indeed experiments show quasicrystals to be very poor electrical conductors. Theoretical calculations of the mean square displacement and the distribution of band velocities in tight binding models [4] show that these systems indeed cannot be described by a standard Boltzmann type theory.

A good way to probe electronic properties in real space consists of introducing impurities, and studying the effects of the perturbation. For noninteracting electrons this leads to indirect exchange interactions between magnetic impurities, and can result in a complex magnetic order [5]. These result from the nontrivial, spatially inhomogeneous response functions for an electron in the quasicrystal. This type of spatial fluctuation leads, for the many body interacting system, to novel behaviors as well. Thus, as a mean-field treatment of the Hubbard model showed [6], inhomogeneous ferromagnetic or antiferromagnetic states can appear for arbitrarily small values of the Hubbard parameter U.

Here we consider the screening of a local moment due to hybridization between a single localized state and the band of conduction electrons in a quasicrystal. In a periodic metallic solid, antiferromagnetic coupling between the conduction electron gas and a localized spin-1/2 leads to the Kondo effect, namely the screening of the moment at low temperatures. In the quasicrystal,

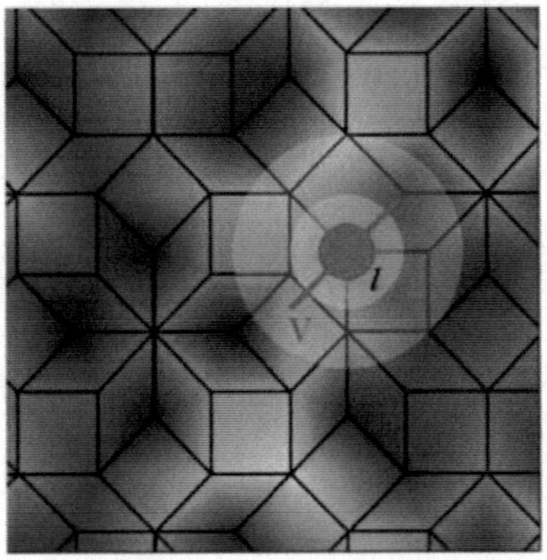

Figure 1.
Schematic drawing of an impurity atom (red) bearing a local moment, located on a given site of a quasiperiodic tiling. The background colors depict the values of the local conduction electron density of states.

one can expect that the screening will be strongly dependent on spatial position of the impurity and on the Fermi energy of the system. This is effectively the case, as was found in [7] for a specific single impurity model. Fig.1 gives a schematic view of the model. An impurity in the quasicrystal background sees a fixed certain local environment and a corresponding local density of states (LDOS). To model the quasicrystalline background, we consider the tight-binding approximation in a quasiperiodic tiling in d dimensions (d=2 or 3), simplified by assuming a single-orbital per site and hopping terms only. The magnetic impurity is described in terms of the infinite U Anderson model, where a single f-orbital is located at a specific site of the quasicrystal. A hybridization term allows for the hopping of electrons between the conduction band and the f-level.

Sec.2 presents the model and Sec.3 the main results for the Kondo temperatures. In Sec.4 the mechanism underlying this form of the distribution is discussed. In Sec.5 temperature-dependence of the average magnetic susceptibility and relation to experimental data in the YbAuAl compounds are discussed.

2. The model

We define the tight binding Hamiltonian for the conduction electrons on a tiling in d dimensions, with a single orbital on each of the vertices. The particles hop along the edges joining the vertices with a hopping amplitude t, assumed to be the same for all edges. The onsite energies are taken to be uniform and are set henceforth equal to zero. The tight-binding Hamiltonian is then

$$\mathcal{H}_c = -t \sum_{\langle ij \rangle, \sigma} \left(c^\dagger_{i\sigma} c_{j\sigma} + c^\dagger_{j\sigma} c_{i\sigma} \right). \tag{1}$$

where $c^\dagger_{i\sigma}$ ($c_{i\sigma}$) is the creation (annihilation) operator of a c-electron at site i with spin σ. The tilings considered for a detailed analysis were square periodic approximants of the Ammann-Beenker (octagonal) tiling in two dimensions (2D) and of the rhombohedral tiling in three dimensions (3D). The most detailed studies were carried out for the 2D system, where it was possible to consider a series of finite size approximants ranging from N = 41 to 8119, where N is the number of sites in the approximant. The 3D system was studied next and served to check the robustness of our results under change of dimensionality.

The Hamiltonian for the conduction electrons is diagonalized assuming twisted boundary conditions, ie, $\psi(x+L, y+L) = e^{i(k_xL+k_yL)}\psi(x,y)$. The densities of states (DOS) are then averaged over a large number of values of $\vec{k}$. This results in a local smoothing of the DOS, and better stability of the numerical calculations. The results thus obtained for the Kondo temperatures are thus valid in a statistical sense. This procedure is also physically pertinent in that it mimics the experimental systems, where finite size effects due to grain boundaries or defects tend to smear out the fine structure in the electronic structure of the quasicrystal.

The impurity is described by the $U \to \infty$ Anderson impurity model or AIM (a discussion of models for local moments in solids can be found in, for example, [8])

$$\mathcal{H} = \mathcal{H}_c + E_f \sum_\sigma n_{f\sigma} + V \sum_\sigma \left(f^\dagger_{\ell\sigma} c_{\ell\sigma} + c^\dagger_{\ell\sigma} f_{\ell\sigma} \right). \tag{2}$$

This model describes a band of non-interacting electrons (c-band) which hybridize with a localized f-orbital located at site ℓ. The operator $f^\dagger_{\ell\sigma}$ ($f_{\ell\sigma}$) creates (destroys) an electron with spin σ at the impurity site ℓ. E_f is the f-level energy, measured with respect to the chemical potential μ, and the hybridization V couples the impurity site to the conduction band. The $U \to \infty$ limit imposes the constraint $n_{f\sigma} = f^\dagger_{\ell\sigma} f_{\ell\sigma} \leq 1$

In terms of the parameters of the AIM, the so-called Kondo coupling is defined by $J = 2V^2/|E_f|$. The Kondo regime corresponds to the limit in which the impurity f-orbital is singly occupied and only spin fluctuations are relevant. The Kondo temperature then gives the temperature below which the impurity moment couples strongly to the conduction electrons, and the moment is effectively quenched.

In order to obtain quantitative results, we consider a large-N (N here represents the quantum number of the spin) limit of Eq. 2 where a variational method can be applied [9]. The two variational parameters are Z_ℓ (quasiparticle weights) and $\tilde{\varepsilon}_{f\ell}$ (renormalized f-energy levels). These are site-dependent in the case of a quasicrystal. The f-level occupation is then given by $n_{f\ell} = 1 - Z_\ell$.

3. Solutions in the Kondo limit

The variational calculation involves placing the impurity site in an effective bath depending on the local environment of the site, ℓ. The hybridization functions $\Delta_{f\ell}$, which gives the broadening of the f-level, are related to the local (onsite) conduction electron Green's functions $G_{c\ell}$ by $\Delta_{f\ell}(\omega) = V^2 G_{c,\ell}(\omega)$. The integral equations involving $\Delta_{f\ell}$ and variational parameters are solved numerically, for each of the sites of the tiling. The Kondo limit is achieved when the impurity site is singly occupied, $n_{f\ell} \approx 1$. The Kondo temperature is, in this limit, given by the width of the resonance at the Fermi level, $T^\ell_{\mathrm{K}} \equiv Z_\ell \mathrm{Im}\left[\Delta_{f\ell}(0)\right]$. The distribution of Kondo temperatures $P(T_{\mathrm{K}})$ can then be calculated. The calculations are repeated for different values of the chemical potential μ and of the parameters V and E_f.

The results show the result for the distribution $P(T_{\mathrm{K}})$ of Kondo temperatures T_K for different locations of the impurity (sample size N=1393, $\mu = -1.8t$, $V = 1.0t$, $E_f - \mu = -3t$). One sees that for small T_K, the distribution follows a power law $P(T_{\mathrm{K}}) \propto T_{\mathrm{K}}^{\alpha-1}$ where the exponent $\alpha < 1$, ie, there is a singular tail for small T_K. It is seen that the singular behavior is obtained for all the values of μ tested, providing that the Kondo coupling J is small enough. The exponent α is non-universal, and depends on the chemical potential and the value of J. In the next section, we will discuss the possible mechanisms underlying this power law.

The wide spread of values of T_{K} in the Kondo regime is reminiscent of the distributions found in disordered metals in the non-Fermi liquid regime [11]. In contrast to disordered metals,

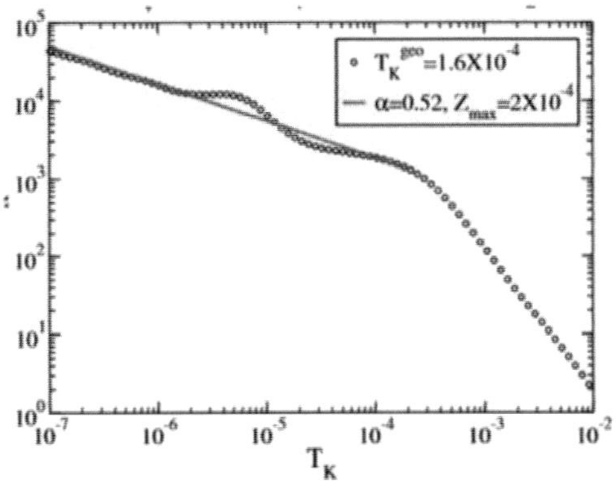

Figure 2.
Distribution function $P(T_\mathrm{K})$ for chemical potential $\mu = 1.8t$ and system size N=1393. The best fit values of the exponent α and of the geometric mean of the Kondo temperatures (defined as in the text, with Z_{max} the upper cutoff in the integral) are given.

however, in the quasicrystal the distribution is *intrinsic*, and does not depend on parameters such as the strength of the disorder. The typical temperature scale in both these situations is measured not by the arithmetic mean, but the geometric mean, $\langle T_\mathrm{K}\rangle$, which is defined by

$$\langle T_\mathrm{K}\rangle = \exp\left(\int dT_\mathrm{K} \log T_\mathrm{K} P(T_\mathrm{K})\right) \tag{3}$$

In the computation the integral runs over the interval of T_K in which P satisfies the power law.

4. Analysis of results

The observed distribution of screening temperatures T_K results from a combination of factors. Clearly, the spread of values of T_K is the result of the inhomogeneous local density of states, $\rho_{c\ell}(E) = (-1/\pi)G_{c\ell}(E)$, in the quasicrystal. The form of the LDOS (see Fig.3) can be qualitatively understood as a signature of multifractal "critical" states. Multifractal properties of electronic states are well-understood in one dimension, where the characteristic exponents can be analytically found in some cases [10]. However, in higher dimensions, it has not been possible to obtain a theoretical description of wavefunctions of the standard tight-binding models, except in some special cases.

To probe the mechanism underlying the distributions $P(T_\mathrm{K})$, let us consider how T_K is related to the electronic properties of the tiling. Although there are no exact solutions of the mean field equations, one can make some simplifying assumptions in the Kondo limit, when E_f lies well below the Fermi level, and $n_{f\ell}$ tends to 1. In this limit, the mean field equations yield an expression of the form

$$\begin{aligned} T_{K\ell} &= D\exp\left[-\frac{\pi}{2}\left(\frac{Re[\Delta_{f\ell}(\mu)] + |E_f - \mu|}{Im[\Delta_{f\ell}(\mu)]}\right)\right] \\ &= D\exp\left[-\frac{1}{2J\rho_{c\ell}(\mu)} + \left(\frac{\mu - Re[\Delta_{c\ell}(\mu)]}{Im[\Delta_{c\ell}(\mu)]}\right)\right] \end{aligned} \tag{4}$$

where the hybridization function was eliminated in favor of the real and imaginary parts of the cavity function $\Delta_{c\ell}$, defined in terms of the electron Green's function by $G_{c\ell}(\omega) = (\omega - \Delta_{c\ell})^{-1}$.

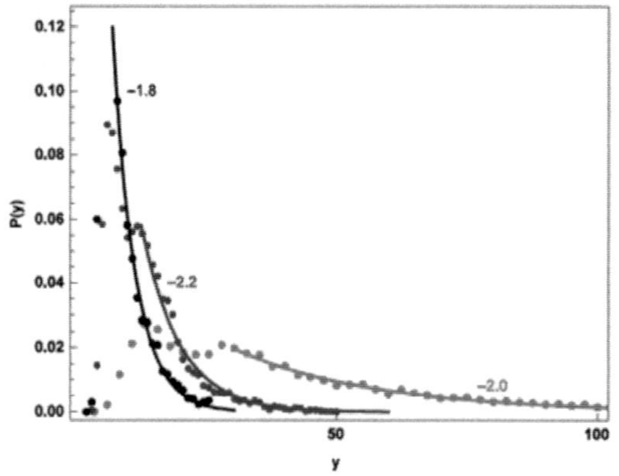

Figure 3.
Distribution functions $W(y)$ of the inverse LDOS y, with fits to an exponential tail for three values of the chemical potential: $\mu = -1.8t$(black), $-2t$(red) and $2.2t$(blue). System size N=1393.

In the Kondo limit, assuming that $J \ll 1$, one can write

$$T_{K\ell}/D = \mathrm{e}^{-\frac{1}{2J}(y_\ell + \mathcal{O}(J))} \tag{5}$$

where the term $\mathcal{O}(J)$ in the exponent can be neglected in the small J limit. The variable $y_\ell = 1/\rho_{c\ell}$ is the inverse LDOS on site ℓ. In this expression for the Kondo temperature, the first term in the exponent dominates for small J, leading to vanishingly small T_{K} on sites with smallest LDOS. To calculate the distribution for small T_{K}, we need the distribution function for large values of y_ℓ. These typically correspond to sites where large fluctuations of the real and imaginary parts of the electron self-energy occur for small energy and boundary condition changes. In consequence, as numerical calculations confirm for the three choices of μ shown in Fig. 3, the y_ℓ behave like Poisson-distributed pseudo-random quantities. The distribution of the variable y (computed for finite broadening of levels and averaged over an energy interval of width 2δ where δ is the mean level spacing) is well-approximated by an exponential form $W(y) = e^{-cy}$. The constant c depends on the nature of the wavefunctions, and thus varies strongly with the position of μ, ranging from 0.05 to 0.2, in the figure. One can now deduce from this form the power law for Kondo temperatures, $P(T_{\mathrm{K}}) \sim T_{\mathrm{K}}^{\alpha-1}$ where $\alpha = 2Jc$.

The exponents α predicted by this argument are non-universal. The constants c vary significantly when the Fermi level is varied. Their values depend slightly on finite size effects and the energy-broadening introduced in the calculation. The smallest value of c occurs for $\mu = -\pm 2t$ i.e. at the main pseudo-gaps of the pure hopping model, where the distribution of y is widest. The exponents for fixed chemical potential are, secondly, proportional to the coupling J. The numerical results for T_{K} indeed show these trends [7] for J small.

The asymptotic form of the distribution is valid in the limit of large size, where distributions of the LDOS and the $\Delta_{f\ell}$ are sufficiently wide. For small approximants, the LDOS distribution will not display an exponential tail, and the minimal value of T_{K}, T_{min}, will not be particularly small. In numerical calculations, the size of the approximant is thus an important limiting factor, and the singular power law dependence is clearly seen only for sufficiently large systems.

IOP Conf. Series: Journal of Physics: Conf. Series **809** (2017) 012021 doi:10.1088/1742-6596/809/1/012021

5. Discussion and conclusions

The singular (power law) form of $P(T_K)$ indicates that, for a finite fraction of sites, quenching of the impurity local moments does not occur, down to very low temperatures. This leads to a non-Fermi-liquid (NFL) behavior as we now argue. Recall that $\chi(T)$ of the Fermi liquid is finite as $T \to 0$, while a free moment shows Curie behavior, ie, has a diverging magnetic susceptibility $\chi(T)$. The zero temperature susceptibility $\chi(T=0)$ in a system with a quenched moment has a finite value, of the order of T_K^{-1}. These behaviors can be compactly written in terms of an interpolative form of the susceptibility for low T, $\chi(T) \sim (T+T_K)^{-1}$. In a simple lattice, T_K is single valued, whereas in disordered metals – or as we have seen, in a quasicrystal – T_K can have different values depending on the local environment.

Let us now consider a quasicrystal with a finite but low concentration of impurities, and let us suppose that one can neglect interactions between them. The experimentally measured susceptibility is the magnetic susceptibility averaged over all sites of the sample, defined by $\overline{\chi}(T) = \int dT_K P(T_K)\chi(T)$. Substituting the form of the distribution for low T_K, a power law extending down to $T_K = 0$ for the infinite quasicrystal, one finds a power law $\overline{\chi}(T) \propto T^{\alpha-1}$, ie, the susceptibility is singular for $\alpha < 1$. Our scenario thus explains the power law of the susceptibility found experimentally as reported in [1, 2]. In this model, the spin contribution to other thermodynamic quantities such as the specific heat should also possess power law singularities. Experiments show however that the singularity of the specific heat is weaker. This could be an indication that impurity-impurity correlations, which were neglected in this simple model, are important.

Small approximants of the quasicrystal should, in contrast, show a Fermi liquid low temperature behavior. For, as the periodic length (and number of local environments) gets smaller, the Kondo temperatures increase, and their lower bound, T_{min}, will rise. For T lower than the value of T_{min}, the susceptibility will thus level off to a constant. This is in accord with experiments which show a nonsingular behavior of the susceptibility in the approximant crystal [1].

To conclude, we have presented an argument for NFL behavior in quasicrystals based on the existence of critical states at the Fermi energy and numerical results that support the scenario. The results are robust under a change of dimension: similar results were found in [7] for a three dimensional tiling. Our model considered properties in the Kondo regime where spin fluctuations are dominant, while theoretical models to study valence fluctuation effects have been investigated elsewhere [12, 13]. More theoretical studies are clearly necessary in order to understand the real-world experimental systems.

References

[1] Deguchi K, Matsukawa S, Sato N K, Hattori T, Ishida K, Takakura H, and Ishimasa T, *Nature Materials* **11**, 1013 (2012)
[2] Watanuki T, Kashimoto S, Kawana D, Yamazaki T, Machida A, Tanaka Y, and Sato T J, *Phys. Rev. B* **86**, 094201 (2012)
[3] Grimm U and Schreiber M, *Quasicrystals - Structure and Physical Properties*, edited by H.-R. Trebin (Wiley-VCH, Weinheim, 2003), pp. 210-235
[4] Trambly de Laissardire G and Mayou D, *C. R. Physique* 15, 70 (2014)
[5] Thiem S and Chalker J, *Europhysics Letters* **110** 17002 (2015)
[6] Jagannathan A and Schulz H J, *Phys. Rev. B* **55** 8045 (1997)
[7] Andrade E C, Jagannathan A, Miranda E, Vojta M and Dobrosavljevic V, *Phys. Rev. Lett.* **115** 036403 (2015)
[8] Hewson A C, *The Kondo Problem to Heavy Fermions*, (Cambridge University Press, Cambridge, 1993), 1st ed.
[9] Coleman P, *Phys. Rev. B* **29** 3035 (1984)
[10] Mace N, Piéchon F and Jagannathan A, *Phys. Rev. B* **93** 205153 (2016)
[11] Tanaskovic D, Miranda E and Dobrosavljevic V, *Phys. Rev. B* **70** 205108 (2004)
[12] Takemura S, Takemori N and Koga A, *Phys. Rev. B* **91** 165114 (2015)

[13] Otsuki J and Kusunose H, J.P.S.J. **85** 073712 (2016)

IOP Conf. Series: Journal of Physics: Conf. Series **809** (2017) 012022 doi:10.1088/1742-6596/809/1/012022

Cluster Size Effects on Electronic Reconstruction in Quasiperiodic System

Ryu Shinzaki, Joji Nasu, and Akihisa Koga
Department of Physics, Tokyo Institute of Technology, Meguro-ku, Tokyo 152-8551, Japan
E-mail: koga@phys.titech.ac.jp

Abstract. We study the extended Anderson model on a two-dimensional Penrose lattice to discuss how the electronic reconstruction in the valence is realized at low temperatures. Making use of the real-space dynamical mean-field theory combined with the continuous-time quantum Monte Carlo method, we find that inhomogeneous electronic distribution appears at low temperatures. The system size dependence is also addressed.

Strong electron correlations in the rare-earth or actinide compounds have attracted great attention, where exotic low-temperature properties have been observed such as heavy fermion behavior [1, 2, 3], valence fluctuations [4, 5], and quantum critical behavior [6]. It has been understood that these interesting properties are attributed to the itinerancy of f electrons through the hybridization to conduction bands and strong electron correlations. Recently, intriguing phenomena have been observed in the Au − Al − Yb alloy [7]. The quasicrystal $Au_{51}Al_{34}Yb_{15}$, where Tsai-type clusters are arranged with quasiperiodic structure, exhibits the divergence in the magnetic susceptibility and the specific heat with unconventional critical exponents. This quantum critical behavior is contrast to heavy fermion behavior in the approximant $Au_{51}Al_{35}Yb_{14}$ with periodic structure. Furthermore, the ytterbium ions have intermediate valence [7, 8, 9], which should suggest that the quasiperiodicity and/or valence fluctuations play an essential role in understanding the exotic behavior of the quasicrystal $Au_{51}Al_{34}Yb_{15}$ [10, 11, 12].

In our previous study, we have considered strongly correlated electron systems on a two-dimensional Penrose lattice [13, 14, 15], making use of the real-space version of dynamical mean-field theory (R-DMFT) [16, 17, 18, 19]. In particular, we have discussed valence fluctuations inherent in the quasiperiodic lattice, by comparing with the results for the periodic lattice. However, it is still unclear whether or not such interesting behavior is realized even in the periodic system with a large unit cell. The first step to answer this fundamental and difficult question is to clarify how the system size affects the electronic reconstruction at low temperatures. To this end, we discuss low-temperature properties in distinct clusters.

We study the extended Anderson lattice model [13, 20, 21, 22, 23], which should be described by the following Hamiltonian,

$$\begin{aligned}\mathcal{H} &= -t\sum_{\langle i,j\rangle,\sigma} c_{i\sigma}^\dagger c_{j\sigma} + V\sum_{i,\sigma}(c_{i\sigma}^\dagger f_{i\sigma} + \mathrm{H.c.}) + \epsilon_f \sum_{i,\sigma} n_{i\sigma}^f \\ &\quad + U_{ff}\sum_i n_{i\uparrow}^f n_{i\downarrow}^f + U_{cf}\sum_{i,\sigma,\sigma'} n_{i\sigma}^c n_{i\sigma'}^f,\end{aligned} \tag{1}$$

IOP Conf. Series: Journal of Physics: Conf. Series **809** (2017) 012022 doi:10.1088/1742-6596/809/1/012022

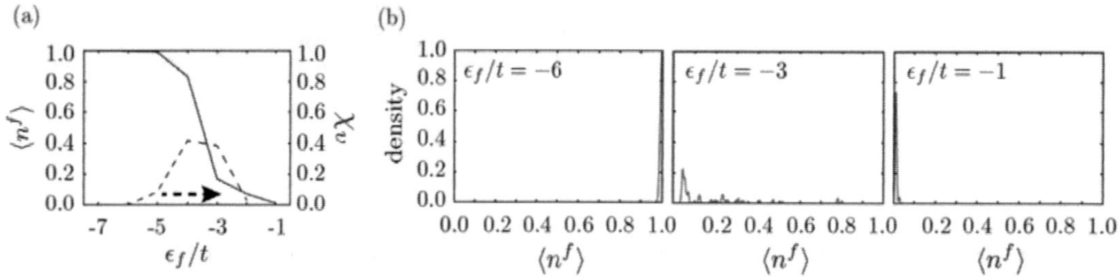

Figure 1. (a) : The f-level dependence of the site-average of the f-electron number (solid line) and valence susceptibility (dashed line) on the lattice with $N = 601$ at $T/t = 0.2$. (b) : The density for each f-electron number in the case of local moment ($\epsilon_f/t = -6$), crossover ($\epsilon_f/t = -3$) and mixed-valence ($\epsilon_f/t = -1$) region on the lattice with $N = 601$ at $T/t = 0.2$.

where $c_{i\sigma}$ ($f_{i\sigma}$) is an annihilation operator of a conduction electron (f electron) with spin $\sigma(=\uparrow,\downarrow)$. $n^c_{i\sigma}(= c^\dagger_{i\sigma}c_{i\sigma})$ and $n^f_{i\sigma}(= f^\dagger_{i\sigma}f_{i\sigma})$ are the number operators of the conduction and f electrons at the ith site, respectively. t is the hopping integral of the conduction electrons between the nearest-neighbor sites, V is the hybridization between the conduction band and f orbitals, and ϵ_f is the energy level of the f orbitals. U_{ff} and U_{cf} are the repulsive interactions between the f electrons, and between the conduction and f electrons. We consider here the vertex model of the two-dimensional Penrose lattice [13, 14, 15]. In this lattice, sites are placed on the corner of the rhombuses and the coordination number Z ranges from three to seven except for the edge sites.

In the paper, we make use of the R-DMFT [16]. In the R-DMFT, the lattice Green's function $\mathbf{G}_\sigma(i\omega_n)$ is assumed to be described by the site-diagonal selfenergy $\Sigma_{i\sigma}(i\omega_n)$. This allows us to introduce the impurity models at each site, in which local electron correlations are taken into account properly. The noninteracting Green's function of an effective bath at the ith site is determined as $\mathcal{G}^{(i)}_\sigma(i\omega_n)^{-1} = [\mathbf{G}^{-1}_\sigma(i\omega_n)]_{ii} + \Sigma_{i\sigma}(i\omega_n)$. These self-consistent calculations are iterated until the Green's functions are converged. In our study, we solve the impurity models by means of the continuous-time quantum Monte Carlo method based on the hybridization expansion [24].

To discuss low temperature properties in the quasiperiodic lattice, we calculate the local quantities such as f-electron number $\langle n^f_i \rangle$ and magnetic susceptibility $\chi_i = \int_0^\beta d\tau \langle M^z_i(\tau) M^z_i \rangle_{\rm imp}$, where $M^z_i = n^c_\uparrow - n^c_\downarrow + n^f_\uparrow - n^f_\downarrow$ is the local magnetic moment operator. For simplicity, the total susceptibility is assumed to be given as $\chi = \sum_i \chi_i/N$, where N is the total number of sites. Here, we use t as the unit of energy. We choose the parameters as $U_{ff}/t = 20$, $U_{cf}/t = 5$, and $V/t = 0.2$. The total electron number per site is fixed as $\langle n^c \rangle + \langle n^f \rangle = 1.9$, where $\langle n^c \rangle \equiv \sum_i \langle n^c_i \rangle /N$, and $\langle n^f \rangle \equiv \sum_i \langle n^f_i \rangle /N$. To discuss how the electronic reconstruction is realized in the quasiperiodic lattice, we treat different clusters with $N = 86$ and 601, which are generated in terms of the inflation-deflation rule [25].

First, we discuss valence fluctuations in the larger cluster with $N = 601$. Figure 1(a) shows the f-level dependence of the valence $\langle n^f_i \rangle$ when $T/t = 0.2$. We find that the increase of ϵ_f decreases the valence from unity to zero without any singularities. Since the $c - f$ spin correlations are not enhanced at the temperature (not shown), this indicates the crossover from the local moment state to mixed-valence state around $\epsilon_f/t \sim -4$, where the valence susceptibility $\chi_v = -\Delta\langle n^f \rangle / \Delta\epsilon_f$ takes a maximum. We wish to note that the inhomogeneous electronic reconstruction appears in the crossover region $\epsilon_f/t \sim -3$ [13, 15], which is characteristic of the

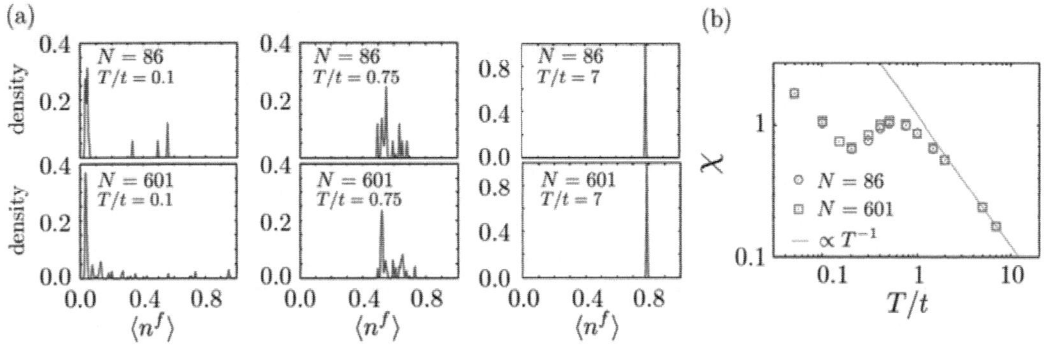

Figure 2. (a) Density profiles for f-electron number in the system with $N = 86$ (upper panel) and $N = 601$ (lower panel) when $\epsilon_f/t = -3$ and $T/t = 0.1, 0.75$, and 7. (b) Bulk magnetic susceptibility in the system with $N = 86$ (the circles) and $N = 601$ (the squares).

quasiperiodic system. Figure 1(b) shows the density profiles for the local moment, crossover, and mixed-valence regions when $\epsilon_f/t = -6$, -3 and -1, respectively. In contrast to the cases of the local moment and mixed-valence regions, the valence distributes in a certain range when the system is in the crossover region. This implies the existence of the electronic reconstruction, which never appears in the periodic systems [13, 15].

To clarify how the electronic reconstruction inherent in a quasiperiodic lattice is realized, we show in Fig. 2(a) the density profiles of the valence at distinct temperatures in the systems with $N = 86$ and 601. At a high temperature $T/t = 7$, the electron valence is uniform in the lattice and $\langle n^f \rangle \sim 0.8$. In the case, we do not find any differences in the results for distinct clusters. Therefore, the quasiperiodic feature does not appear in the correlated electron systems at higher temperatures. However, the decrease of the temperature enhances site-dependent behavior, yielding the valence distribution. When $T/t = 0.75$, $\langle n_i^f \rangle$ distributes in a range $0.5 < \langle n^f \rangle < 0.7$. An important point is that the distributions for the clusters with $N = 86$ and 601 are slightly different from each other. This implies that the correlation length becomes longer with decrease of temperature, and thereby a large cluster, at least, with $N = 601$ sites is necessary to describe the electronic reconstruction correctly. Moreover, at lower temperatures, this behavior becomes more serious. In fact, the valence distribution in a larger cluster is quite different from the smaller one except for the peak around $\langle n^f \rangle \sim 0$. Therefore, a larger cluster may be necessary to describe the electronic reconstruction at lower temperatures more precisely.

In the Penrose lattice, sites are not equivalent to each other, but any cluster exists with a finite density, which is in contrast to the periodic system. Therefore, if the correlation length becomes longer with decreasing temperatures, the valence distributions are drastically changed [13]. In the sense, it should be hard to distinguish the properties of the periodic and quasiperiodic systems when the correlation length is shorter than the scale of its unit cell.

Finally, we consider how the magnetic response is affected by the system size and boundary effects. Figure 2 (b) shows the temperature dependence of the magnetic susceptibility χ. In this parameter with $\epsilon_f/t = -3$, nonmonotonic behavior appears in the magnetic susceptibility, which is mainly reflected by the change in the number of valence [15]. Although the valence distribution strongly depends on the system size discussed above, the bulk magnetic susceptibility has little changed. This may imply that it is hard to see interesting electronic properties reflected by the quasiperiodic nature discussed above in terms of the bulk quantities.

In our work, only two different clusters with global fivefold rotational symmetry have been dealt with due to numerical limitations. Although the systematic analysis for the cluster size, cluster shape, and/or boundary dependence have not been done, we have carefully examined local and bulk quantities in distinct clusters. We have then clarified that the cluster dependence becomes crucial with decreasing temperatures. This implies that, in the system with large valence fluctuations, the correlation length plays an important role and the effect of the cluster size and/or its shape should not be negligible at low temperatures. We have confirmed that in the Kondo region, the cluster size little affects low temperature properties (not shown) since the Kondo singlet is stabilized locally and the correlation length is very short. This is common to the physics in the strongly correlated electron systems with the periodic lattice. Therefore, low temperature properties characteristic of the quasiperiodic structure should appear when spin, orbital and/or valence fluctuations are enhanced.

In summary, we have investigated the extended Anderson model on the two dimensional Penrose lattice. Making use of the R-DMFT, we have calculated the f-electron number and bulk magnetic susceptibility. By comparing with the results with distinct clusters, it has been clarified that the correlation length becomes longer with decreasing temperatures and the electronic reconstruction appears inherent in the Penrose lattice.

Acknowledgments

We would like to thank N. Takemori for valuable discussions. Parts of the numerical calculations were performed in the supercomputing systems in ISSP, the University of Tokyo. This work was partly supported by the Grant-in-Aid for Scientific Research from JSPS, KAKENHI Grant Number 25800193, 16H01066 (A.K.), and 16K17747 (J.N.). The simulations have been performed using some of the ALPS libraries [26].

References

[1] Andres K, Graebner E J, and Ott R H: Phys. Rev. Lett. **35** (1975) 1779.
[2] Stewart R G: Rev. Mod. Phys. **56** (1984) 755.
[3] Satoh K, Fujita T, Maeno Y, Onuki Y, and Komatsubara T: J. Phys. Soc. Jpn. **58** (1989) 1012.
[4] Gschneidner A K and Eyring L: *Handbook on the Physics and Chemistry of Rare Earths* (1978)
[5] Manheimer A M and Parks D R: Phys. Rev. Lett. **42** (1979) 321.
[6] Doniach S: Physica B+C **91** (1977) 231.
[7] Deguchi K, Matsukawa S, Sato K N, Hattori T, Ishida K, Takakura H, and Ishimasa T: Nat. Mat. **11** (2012) 1013.
[8] Matsukawa S *et al.*: J. Phys. Soc. Jpn. **83** (2014) 034705.
[9] Watanuki T, Kashimoto S, Kawana D, Yamazaki T, Machida A, Tanaka Y, and Sato J T: Phys. Rev. B **86** (2012) 094201.
[10] Watanabe S and Miyake K: J. Phys. Soc. Jpn. **82** (2013) 083704.
[11] Andrade C E, Jagannathan A, Miranda E, Vojta M, and Dobrosavljević: Phys. Rev. Lett. **115** (2015) 036403.
[12] Otsuki J, Kusunose H: J. Phys. Soc. Jpn. **85** (2016) 073712.
[13] Takemura S, Takemori N, and Koga A: Phys. Rev. B **91** (2015) 165114.
[14] Takemori S and Koga A: J. Phys. Soc. Jpn. **84** (2015) 023701.
[15] Shinzaki R, Nasu J, and Koga A: J. Phys. Soc. Jpn. **85** (2016) 114706.
[16] Georges A, Kotliar G, Krauth W, and Rozenberg J M: Rev. Mod. Phys. **68** (1996) 13.
[17] Helmes R W, Costi T A, and Rosch A: Phys. Rev. Lett. **100**, 056403 (2008).
[18] Snoek M, Titvinidze I, Töke C, Byczuk K, and Hofstetter W: New J. Phys. **10**, 093008 (2008).
[19] Koga A, Higashiyama T, Inaba K, Suga S, and Kawakami N, J. Phys. Soc. Jpn. **77**, 073602 (2008); Phys. Rev. A **79**, 013607 (2009).
[20] Onishi Y and Miyake K: J. Phys. Soc. Jpn. **69**, 3955 (2000).
[21] Watanabe S, Imada M, and Miyake K: J. Phys. Soc. Jpn. **75** (2006) 043710.
[22] Kojima Y and Koga A: JPS Conf. Proc. **1** (2013) 012106.
[23] Shinzaki R, Nasu J, and Koga A: J. Phys.: Conf. Ser. **683** (2016) 012041.
[24] Werner P, Comanac A, de' Medici L, Troyer M, and Millis J A: Phys. Rev. Lett. **97** (2006) 076405.
[25] Levine D and Steinhardt P J: Phys. Rev. Lett. **53**, 2477 (1984).
[26] Bauer B *et al.*: J. Stat. Mech. Theory Exp. **2011** (2011) P05001.

Gap structure of 1D cut and project Hamiltonians.

Nicolas Macé, Anuradha Jagannathan and Frédéric Piéchon

Laboratoire de Physique des Solides, Université Paris-Saclay, 91400 Orsay, France

E-mail: nicolas.mace@u-psud.fr

Abstract. We study the gap properties of nearest neighbors tight binding models on quasiperiodic chains. We argue that two kind of gaps should be distinguished: stable and transient. We show that stable gaps have a well defined quasiperiodic limit. We also show that there is a direct relation between the gap size and the gap label.

Introduction

The gap labeling theorem (GLT) [1] is one of the few exact results known about electronic properties of quasicrystals. It concerns the *integrated density of states* (IDOS), which is defined as $\text{ids}(E)$ = fraction of energy states below energy E. Inside a spectral gap, the IDOS is of course constant, and the theorem constrains the value it can take. In the case of canonical cut and project chains, the theorem states that

$$\text{idos}(E \in \text{gap}) = \frac{n}{1+\alpha} \mod 1 = \frac{n}{1+\alpha} + m \tag{1}$$

where α is the irrational involved in the construction of the quasiperiodic chain. m and n are integers, n is the label of the gap, and $m(n)$ is such that the IDOS satisfies $0 \leq \text{idos} \leq 1$.

The GLT can appear mysterious and profound. However, it has a simple derivation, making use of the approximants. Let $\alpha_l = p_l/q_l$ be a convergent of α. Consider the periodic approximant built with α_l. Its spectrum consists in general of $N_l = p_l + q_l$ energy bands [2, 3], and thus the IDOS inside a gap is given by

$$\text{idos}(E \in \text{gap}) = \frac{j(E)}{N_l} \tag{2}$$

where $j(E)$ counts the number of bands below energy E. Since p_l and q_l are relatively prime, using Bézout's identity we can write $j = ap_l + bq_l = (a-b)q_l + bN_l$. We let $n = a - b$ and we remark that

$$\text{idos}(E \in \text{gap}) = n\frac{q_l}{N_l} \mod 1 \tag{3}$$

Taking the quasiperiodic limit $l \to \infty$ at fixed n, we recover the GLT. Moreover, the formula (3) extends the GLT to approximants, which, as we will see, provides useful insight into the gap structure.

As such, the theorem contains little information: it only constrains the set of values the IDOS can take inside the gaps, but does not tell anything about the energy location of the gaps, or about their width. Nevertheless, as we shall see, the gap label n contains additional information the GLT does not explicitly tell us about.

1. Nearest neighbors tight-binding model on quasiperiodic chains

We consider an electron hopping on a quasiperiodic chain, obtained by the canonical cut and project method [4]. As a generic example, we consider a model where the nearest neighbors hopping amplitude takes two values according to the quasiperiodic sequence:

$$H(\alpha) = \sum_{x \in \mathbf{Z}} t_{x,x+1} |x\rangle \langle x+1| + \text{h.c.} \tag{4}$$

where $t_{x,x+1} = t_1$ or $t_{x,x+1} = t_2$. Once the irrational is fixed, the only free parameter in the model is the ratio of the two jump amplitudes $\rho = t_1/t_2$.

This model can be realized experimentally with light [5] or cavity polaritons [6]. In the latter case, the details of the energy spectrum could be accurately measured, leading the authors to observe subtle phenomena such as the energy displacement of an edge state within a spectral gap. Such observation makes theoretical studies of the spectral gaps particularly relevant.

2. Application: the Fibonacci quasicrystal

Figure 1: The gap labels of the first few Fibonacci approximants. Blue: stable gaps, red: transient gaps.

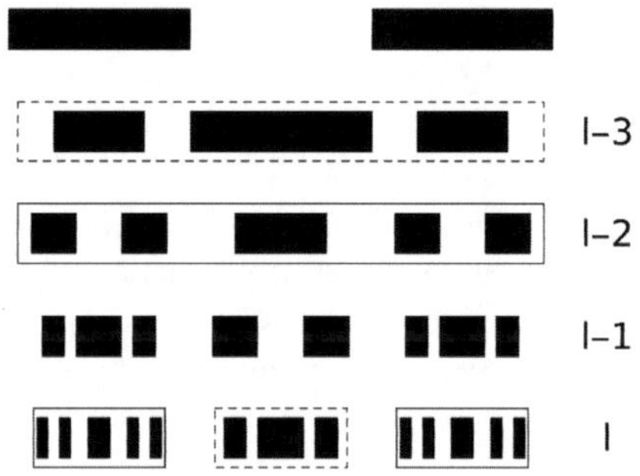

Figure 2: The a recursive construction of the Fibonacci spectrum.

Energy spectra of the first few Fibonacci approximants. Here we took $\rho = 0.5$, and the spectra were approximated using the renormalization group method [7, 8].

We now examine in detail the special case of the Fibonacci chain, which will give us more insights. It is characterized by $\alpha = \omega$, where $\omega = (\sqrt{5}-1)/2$ is the inverse of the golden ratio. The gap labeling is thus

$$\text{idos}(E \in \text{gap}) = n\omega \mod 1 \tag{5}$$

The approximants are built with consecutive Fibonacci numbers: $p_l = F_{l-2}$, $q_l = F_{l-1}$. The lth approximant has $N_l - 1 = F_l - 1$ gaps. Naively, we could expect all these gaps to remain open in the quasiperiodic limit $l \to \infty$. This is not true. For example, the approximants of the form $l = 3m$ have a gap at $E = 0$ (fig (1)), but it closes in the quasiperiodic limit, as can be seen directly on the approximants, or by the fact that there is no n such that $n\omega \mod 1 = 1/2$. We will say that such a gap is *transient*. On the contrary, the two largest gaps of the Fibonacci approximants (labeled $n = \pm 1$ see fig (1)) appear to be *stable*: they stay open all the way to the quasiperiodic limit.

The behavior of an electron of the Fibonacci chain is well understood. One can show [7, 8] that the spectrum of the lth approximant can be approximated by a simple linear combination of the spectra of the approximants $l-2$ and $l-3$ (fig (2)). This provides us with a recursive gap labeling procedure. Let $\mathbf{g} = (m, n)$ be the vector containing the labels of a gap (1). Let G_l be the set of all gap labels $\mathbf{g}$ of the lth approximant. Using the recursive spectrum construction of figure (2) we see that G_l can be partitioned into 3 subsets containing the gap labels of the left, central and right clusters of energy bands. We have the following recursive relations [9]:

$$G_l^{\text{left}} = M^{-2} G_{l-2} \tag{6}$$
$$G_l^{0} = M^{-3} G_{l-3} + \mathbf{g}_1 \tag{7}$$
$$G_l^{\text{right}} = M^{-2} G_{l-2} + \mathbf{g}_2 \tag{8}$$

where $\mathbf{g}_1 = (1, -1)$ and $\mathbf{g}_2 = (0, 1)$ are the labels of the two main gaps (corresponding to $n = \pm 1$, see figure (1)), and M is the geometrical substitution matrix

$$M = \begin{bmatrix} 1 & 1 \\ 1 & 0 \end{bmatrix} \tag{9}$$

that can be used to generate the Fibonacci sequence.

This recursive relation provides us with a very simple geometrical interpretation of the GLT, in the simple case of the Fibonacci chain. Moreover, the construction shows that the transient gaps are the iterates of the $E = 0$ gap, while the stable gaps are the iterates of the two main gaps.

This example shows that the GLT can be extended to the approximants, at the price of introducing transient gaps that are not present in the spectrum of the quasiperiodic system. Fortunately, we are able to pinpoint them. We now turn to the general characterization of the transient and stable gaps in the case of an arbitrary cut and project chain.

3. Transient and stable gaps

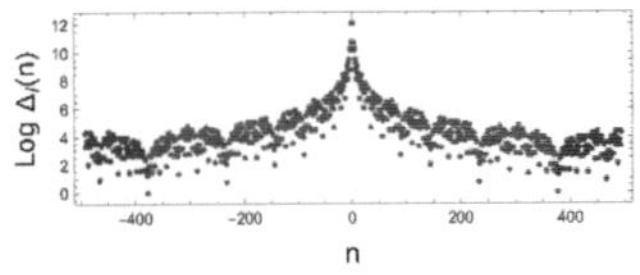

Figure 3: Fibonacci chain of $N_{16} = 987$ atoms.

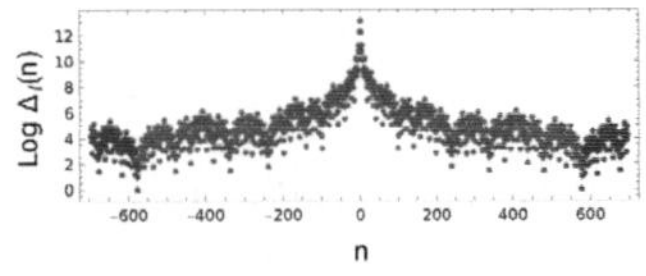

Figure 4: Silver mean chain of $N_{10} = 1393$ atoms.

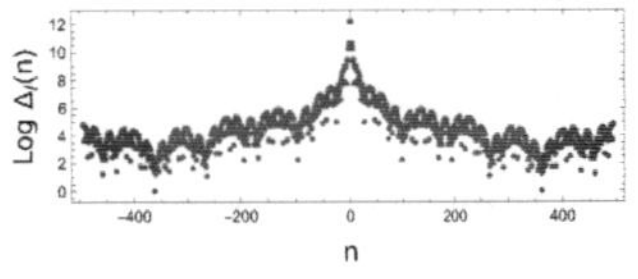

Figure 5: Chain $\alpha = \sqrt{3} - 1$ of $N_{12} = 989$ atoms.

Gap width as a function of gap label for several quasiperiodic chains. Blue: stable gap, red: transient gaps, according to criterion (11). The ratio of the couplings was set arbitrarily to $\rho = 0.5$.

In the previous part, we considered a gap centered around energy $\overline{E}$ to be transient if its width $\Delta_l(\overline{E})$ vanished:

$$\Delta_l(\overline{E}) \underset{l \to \infty}{\longrightarrow} 0 \tag{10}$$

We introduce here a more practical criterion. Let $\overline{E}_l(n)$ be the mean energy inside the gap of label n, for the lth approximant. Define the displacement of a gap as $D_l(n) = |\overline{E}_{l+1}(n) - \overline{E}_l(n)|$ We say the gap n at step l is transient if its half-width does not overlap significantly its counterpart at step $l + 1$, ie if

$$a(\rho)(\Delta_l(n) + \Delta_{l+1}(n)) < D_l(n), \tag{11}$$

and stable otherwise. We have introduced the *overlap coefficient* $a(\rho)$, which quantify how much overlap we require for a gap to be considered stable. Numerically, we determine a by requiring that any stable gap at step l overlaps their counterpart at step $l + L$ with L arbitrarily large. As the ratio of the couplings ρ is decreased, gaps are enlarged and overlap more and more. Thus, the value of a also decreases with ρ (we thank the referee for pointing out that fact). As an example, we find $a(0.5) = 0.25$ for the Fibonacci chain.

It is easy to see that criterion (11), applied to the Fibonacci chain, yields the same stable and transient gaps as criterion (10) It is not trivial that these two criteria are in general equivalent. However, it does appear to be the case for all the quasiperiodic chains we tested numerically. We will therefore consider them equivalent from now.

Figures (3) (4) (5) show the widths of the gaps as a function of their label, for various quasiperiodic chains. The global structure is similar in every case: the gap width is a globally decreasing function of the gap index (this is consistent with perturbation theory in the limit $\rho \to 1$, see eg. [10]). Here we also observe that width decreases as a power-law of the index. In the case of the Fibonacci chain, we notice log-periodic oscillations, stemming from the scale invariance of the Fibonacci Hamiltonian. In all cases, there is a critical label above which all gaps are transient, and below which all gaps are stable.

Conclusion and perspectives

We have shown that the well known GLT is applicable not just to quasiperiodic chains but also to their approximants. The price to pay is the introduction of *transient gaps*, which vanish in the quasiperiodic limit. Fortunately, these gaps are easy to spot: they always have the highest gap label. Moreover, we show that the gap label contains relevant physical information: it globally orders the gaps by decreasing width, and separate stable from transient gaps.

The gap label, which had *a priori* no physical content, thus appears physically relevant. It has also been shown the gap label can be considered as a topological number, and used to characterize edge states [11].

Open problems include more finely describing the behavior of the gap width with the index and extending this description to higher dimensional quasicrystals, which are also known to have gaps [12, 13].

References

[1] Bellissard, J. Ed. by Luck, J., Moussa, P., and Waldschmidt, M. Springer, 1993. Chap. Gap Labeling Theorems for Schrödinger's Operators.

[2] Luck, J. M. In: *Phys. Rev. B* 39 (9 Mar. 1989), pp. 5834–5849.

[3] Sire, C. and Mosseri, R. In: *J. Phys. France* 50.24 (1989), pp. 3447–3461.

[4] Baake, M. and Grimm, U. Cambridge university press, 2013. Chap. Projection Method and Model Sets.

[5] Gellermann, W. et al. In: *Phys. Rev. Lett.* 72 (5 Jan. 1994), pp. 633–636.

[6] Tanese, D. et al. In: *Phys. Rev. Lett.* 112 (14 Apr. 2014), p. 146404.

[7] Niu, Q. and Nori, F. In: *Phys. Rev. B* 42 (16 Dec. 1990), pp. 10329–10341.

[8] Piéchon, F., Benakli, M., and Jagannathan, A. In: *Phys. Rev. Lett.* 74 (26 June 1995), pp. 5248–5251.

[9] Macé, N., Jagannathan, A., and Piéchon, F. In: *Phys. Rev. B* 93 (20 May 2016), p. 205153.

[10] Luck, J. M. In: *Phys. Rev. B* 39 (9 Mar. 1989), pp. 5834–5849.

[11] Levy, E. et al. In: (2015), pp. 1–9. arXiv: `arXiv:1509.04028v2`.

[12] Zijlstra, E. In: *Journal of Non-Crystalline Solids* 334–335 (2004). 8th International Conference on Quasicrystals, pp. 126–129.

[13] Prunelé, E. de. In: *Phys. Rev. B* 66 (9 Sept. 2002), p. 094202.

Brick tiling

S I Ben-Abraham[1] and D Flom[2]

[1] Department of Physics, Ben-Gurion University of the Negev, Beer Sheba, Israel
E-mail: shelomo.benabraham@gmail.com
[2] Department of Electro-Optical Engineering, Ben-Gurion University of the Negev, Beer Sheba, Israel
E-mail: dvir.flom@gmail.com

Abstract. We propose a generalization of the well-known two-dimensional table tiling to three dimensions – the brick tiling. We develop a color/digit code for this tiling. We also discuss possible implications for higher dimensions.

1. Introduction

The work reported in this paper is part of a project to generalize standard one-dimensional sequences and two-dimensional (2D) structures to higher dimensions [1-4]. Here we consider the well-known two-dimensional *table tiling* and put forward its generalization to three dimensions which we call the *brick tiling*. The table tiling was thoroughly discussed by E Arthur Robinson Jr [5]. The interested reader will find wider and deeper mathematical background in the important paper by Boris Solomyak [6] and the seminal book by Jean-Paul Allouche and Jeffrey Shallit [7] and that by Michael Baake and Uwe Grimm [8].

.

2. Preliminaries

A brick tiling (including the table tiling) in any dimension (if it exists) is a *lattice substitution system* [9]. That means that it is supported by a lattice $\mathbb{Z}^d$. Its *prototiles* are *lattice animals* (alias *d-polyominoes,* i.e. d-dimensional polyominoes or d-polycubes) that is connected unions of d-dimensional cubic unit cells (d-cubes) [10]. A 2D brick, for instance, consists of two squares.

In d dimensions, a *standard brick* $\mathbf{B}_d$ (in what follows, simply *brick*) is a d-dimensional cuboid with edges of $2^0 = 1, 2^1 = 2, \dots, 2^{d-1}$ units; hence it consists of $2^{\sum_{k=0}^{d-1} k}$ unit cubes $\mathbf{q}_d$. When there can be no confusion we omit the subscript d. We focus here on three dimensions (3D). There, the prototiles are $\mathbf{B}_3$'s, i.e. 1×2×4 cuboids (octominoes). In what follows we shall call them *protobricks*. A g times inflated brick will be denoted $\mathbf{B}_d(g)$. Thus, $\mathbf{B}_d(0) \equiv \mathbf{B}_d$ is a protobrick.

It is easy to see that in d dimensions the protobricks come in $d!$ orientations; for convenience we shall call them *colors*.

In any dimension *d*, a brick is a *rep-tile*. Paraphrasing Robinson [5] we say:
A *d*-dimensional *rep-tile* (*d-rep-tile,* for short) is a polytope that can be tiled by a finite number of smaller, congruent copies of itself. Obviously, the process can be repeated *ad infinitum.* By the same token, a rep-tile produces, by *inflation,* an infinite tiling.

3. Two dimensions – table tiling

To keep this paper self-contained we briefly recapitulate the essential features of the table tiling (alias *domino tiling*). In the present context we may call it *2D brick tiling.* Fig. 1 (adapted from Robinson [3]) shows its inflation and labeling. Clearly, there are $2! = 2$ protobricks $\mathbf{B}_2$ differing in orientation; each $\mathbf{B}_2$ consists of $2^{d-1} = 2^1 = 2$ unit squares $\mathbf{q}_2$.

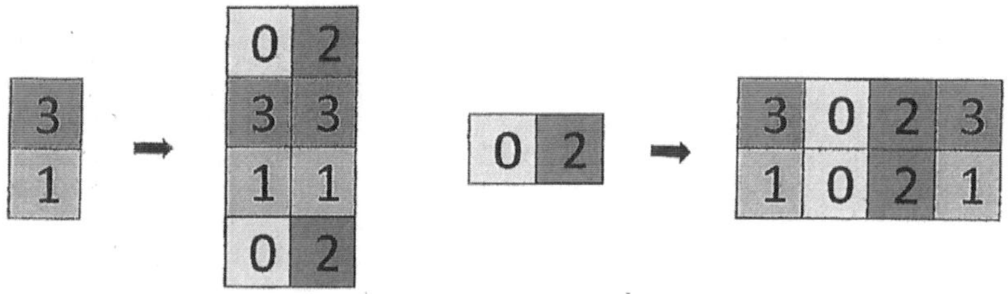

Figure 1. The table tiling inflation and the labeling of its protobricks.

The substitution associated with the inflation is (cf. [5], [8])

$$0 \to \begin{bmatrix} 3 & 0 \\ 1 & 0 \end{bmatrix} \quad 1 \to \begin{bmatrix} 1 & 1 \\ 0 & 2 \end{bmatrix} \quad 2 \to \begin{bmatrix} 2 & 3 \\ 2 & 1 \end{bmatrix} \quad 3 \to \begin{bmatrix} 0 & 2 \\ 3 & 3 \end{bmatrix}. \qquad (1)$$

3. Three dimensions – brick tiling

A consistent generalization of the table tiling must comply with the following conditions:

(a) the inflated brick $\mathbf{B}_d(1)$ consists of 2^d protobricks $\mathbf{B}_d$; in other words, the inflation factor is 2^d;
(b) all bricks are mirror symmetric w.r.t. to their median planes;
(c) there is maximum contact between the protobricks $\mathbf{B}_d$ within the inflated brick $\mathbf{B}_d(1)$;
(d) the inflated brick contains a maximum of colors subject to condition (c).

In 3D there are $3! = 6$ protobricks characterized by their orientation w.r.t. Cartesian coordinate axes x, y, z and briefly labeled by digits and/or colors. They are explicitly listed in Eq. (2) and shown in Fig. 2 along with generation 1 of their inflation. Here the first entry in the parentheses refers to the direction of the long edge, the second to the direction of the medium edge and the third to the direction of the short edge. It can be seen that, indeed, the inflated bricks $\mathbf{B}_3(1)$ (a) consist of $2^3 = 8$ protobricks, (b) are mirror symmetric w.r.t. the median planes, (c) satisfy maximum contact between the proto-bricks $\mathbf{B}_3$, and (d) contain three colors. Four colors would violate the symmetry condition.

The six protobricks of the 3D brick tiling and their inflation are shown in Fig. 2. It is easy to see that the starting orientation is an invariant of the inflation. While the structure would eventually in the limit fill the entire lattice $\mathbb{Z}^3$ it will stay anorthotropic and keep the original shape. Fig. 3 shows the hull of the second generation $\mathbf{B}_3(2)$ starting with $\mathbf{B}_3(0) = 0$ (yellow). The second generation contains $8^2 = 2^6 = 64$ protobricks and hence $8^3 = 2^9 = 512$ unit cubes. Likewise, the third generation contains $8^3 = 2^9 = 512$ protobricks and hence $8^4 = 2^{12} = 4096$ unit cubes. The numbers rapidly become staggering.

$$\begin{aligned} &0 \text{ (yellow)} = (xyz) \quad , \quad 1 \text{ (green)} = (xzy) \,■, \\ &2 \text{ (orange)} = (yzx) \,■, \quad 3 \text{ (blue)} = (yxz) \,■, \\ &4 \text{ (red)} = (zxy) \,■, \quad 5 \text{ (violet)} = (zyx) \,■ . \end{aligned} \qquad (2)$$

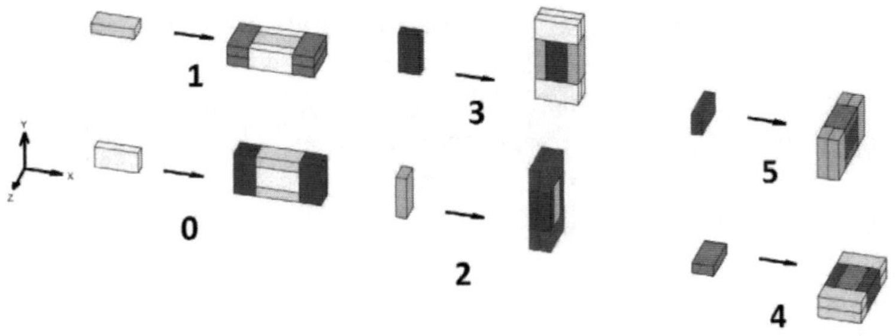

Figure 2. The six protobricks of the 3D brick tiling and their inflation.

Figure 3. Hull of the second generation of the three-dimensional brick tiling starting with orientation 0 (yellow).

4. Arbitrary dimensions

What can be said in the present context about arbitrary dimensions? Bricks exist in all dimensions. A brick tiling fulfilling conditions (a)–(d) certainly exists in one, two and three dimensions. The 1D case is ridiculously degenerate but perfectly valid. Trying to construct a 4D brick tiling we ran into contradictions. One may, though, extend the 2D solution to 3D and even higher dimensions by leaving the shorter edges idle. In other words, one may construct the 2D table tiling using real dominoes. In an analogous way one may extend the 3D brick tiling. This, of course, is not what we really intend. Perhaps, by somewhat relaxing the conditions (a)-(d) one might succeed to generalize the brick tiling to higher dimensions. That remains to be seen.

5. Concluding remarks

The 2D table tiling and its code are quite simple and they seem to be related to the 2D chair tiling and its code be some kind of duality. However, in contrast to the chair tiling whose Fourier spectrum is pure point that of the table tiling already in 2D contains a singular continuous part. It is reasonable to conjecture that that is true in 3D as well. It is altogether rather clear that the 3D brick tiling is much more complex and very different from the 3D chair tiling.

In analogy to 2D we also developed a lattice substitution tiling color coding for the 3D brick tiling. This is reported in our companion paper [11] in this volume.

References

[1] Ben-Abraham S I and Quandt D 2011 Aperiodic structures and notions of order and disorder *Philos. Mag.* **91** 2718-2727
[2] Ben-Abraham S I, Quandt A and Shapira D 2013 Multidimensional paperfolding systems *Acta Cryst.* **A69** 123-130
[3] Ben-Abraham S I, Flom D, Richman R A and Shapira D 2014 Multidimensional "paperfolding" structures – three and four dimensions *Acta Phys. Pol. A* **126** 435-437
[4] Lee J.-Y., Flom D. and Ben-Abraham S I, 2016 Multidimensional period doubling structures *Acta Cryst.* **A72** 391-394
[5] Robinson E A Jr 1999 On the table and the chair *Indag. Mathem. N.S.* **10** 581-599
[6] Solomyak B 1996 Dynamics of self-similar tilings *Ergodic Th. Dynam. Syst.* **17** 695-738
[7] Allouche J-P and Shallit J 2003 *Automatic Sequences: Theory, Applications, Generalizations* (Cambridge University Press)
[8] Baake M.and Grimm U 2013 *Aperiodic Order Volume 1: A Mathematical Invitation* (Cambridge University Press)
[9] Lee J.-Y. and Moody R V 2001 Lattice substitution systems and model sets *Discrete Comput. Geom.* **25** 173-201
[10] Golomb S W 1996 *Polyominoes: Puzzles, Patterns, Problems, and Packings* 2nd ed (Princeton University Press)
[11] Flom D and Ben-Abraham SI 2016 Color coding for the brick tiling *J. Phys. Conf. Series* (this volume)

Color coding for the brick tiling

D Flom[1] **and S I Ben-Abraham**[2]

[1]Department of Electro-Optical Engineering,
Ben-Gurion University of the Negev, Beer Sheba, Israel
E-mail: dvir.flom@gmail.com
[2]Department of Physics, Ben-Gurion University of the Negev,
Beer Sheba, Israel
E-mail: shelomo.benabraham@gmail.com

Abstract. We construct a lattice substitution tiling associated with and coding for the three-dimensional brick tiling. Its prototiles are 48 unit cubes labeled by digits nicknamed “colors”.

1. Introduction

This is a companion and sequel to our paper [1] where we introduced the *brick tiling* – a three-dimensional (3D) generalization of the two-dimensional (2D) *table tiling*. We follow the coding for the two-dimensional (2D) table tiling invented by Robinson [2]; this is shown, with minor adaptations, in Fig.1.

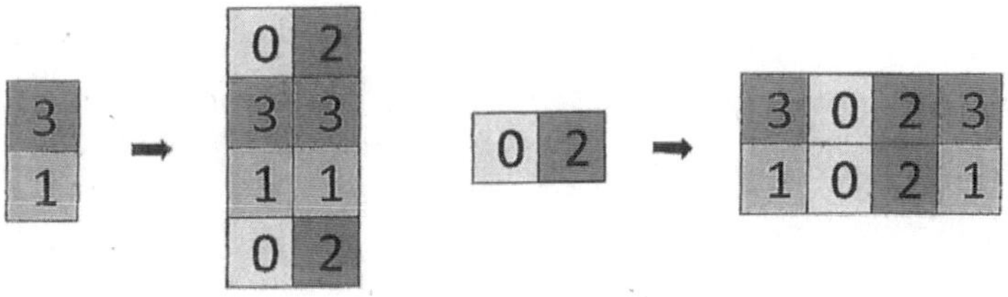

Figure 1. The table tiling inflation and the labeling of its protobricks by digits and colors.

The 2D table tiling and its code are quite simple and they seem to be related to the 2D chair tiling and its code by some kind of duality. It turns out that the 3D brick tiling is much more complex and very different from the 3D chair tiling.

2. Color code substitution

In order to construct the color coding tiling for the 3D brick tiling we label its 48 basic cubes ($\mathbf{q}_3$'s). While in 2D four digits (or colors) suffice, in 3D we would need 48 digits (or colors). This is clearly rather cumbersome and it is practically impossible to find 48 distinguishable hues. Therefore we do the labeling in two stages. One might try a two-color scheme. However, we find it most expedient and intuitive to label the proto-bricks ($\mathbf{B}_3$'s) by six digits 0 through 5 and correspondingly by six colors and label the unit cubes ($\mathbf{q}_3$'s) by eight digits 0 through 7. Thus the unit cubes are, somewhat redundantly, labeled by two digits, the first denoting the brick, the second the cube; for instance [03] denotes cube 3 of brick 0 (yellow). In this way we produce a faithful and maximal coding. This is shown in Fig. 2.

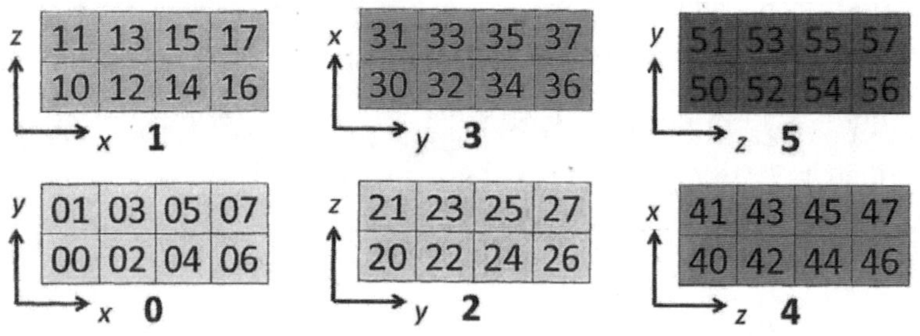

Figure 2. The six protobricks labeled by two digits.

We derive the substitution of the 3D digit/color coding tiling in a way analogous to the procedure proposed by Robinson [2] and clear from Fig. 1. However, in the present case the procedure is somewhat more involved. In the first stage we label the protobricks (**B**$_3$(0)'s) in the first inflated brick (**B**$_3$(1)) by six triples (*pqr*):) (013), (235), (451), (104), (321), (542) as shown in Fig. 3 and construct the substitution shown in Eq. (1). This substitution conserves the orientation of the respective **B**$_3$(0). In a second stage the entire substitution must be transformed to a single orientation. In the present case we chose the orientation (xy) of brick 0 (yellow). The final result is the substitution shown explicitly in Eq. (2).

$$\begin{array}{llll}
[p0]\to\begin{matrix} & \begin{bmatrix}r2&r3\\r0&r1\end{bmatrix}\\ \begin{bmatrix}r2&r3\\r0&r1\end{bmatrix} & \end{matrix} &
[p1]\to\begin{matrix} & \begin{bmatrix}r6&r7\\r4&r5\end{bmatrix}\\ \begin{bmatrix}r6&r7\\r4&r5\end{bmatrix} & \end{matrix} &
[p2]\to\begin{matrix} & \begin{bmatrix}p0&p2\\q1&q3\end{bmatrix}\\ \begin{bmatrix}p0&p2\\q0&q2\end{bmatrix} & \end{matrix} &
[p3]\to\begin{matrix} & \begin{bmatrix}q1&q3\\p1&p3\end{bmatrix}\\ \begin{bmatrix}q0&q2\\p1&p3\end{bmatrix} & \end{matrix} \\
[p4]\to\begin{matrix} & \begin{bmatrix}p4&p6\\q5&q7\end{bmatrix}\\ \begin{bmatrix}p4&p6\\q4&q6\end{bmatrix} & \end{matrix} &
[p5]\to\begin{matrix} & \begin{bmatrix}q5&q7\\p5&p7\end{bmatrix}\\ \begin{bmatrix}q4&q6\\p5&p7\end{bmatrix} & \end{matrix} &
[p6]\to\begin{matrix} & \begin{bmatrix}r2&r3\\r0&r1\end{bmatrix}\\ \begin{bmatrix}r2&r3\\r0&r1\end{bmatrix} & \end{matrix} &
[p7]\to\begin{matrix} & \begin{bmatrix}r6&r7\\r4&r5\end{bmatrix}\\ \begin{bmatrix}r6&r7\\r4&r5\end{bmatrix} & \end{matrix}
\end{array} \quad (1)$$

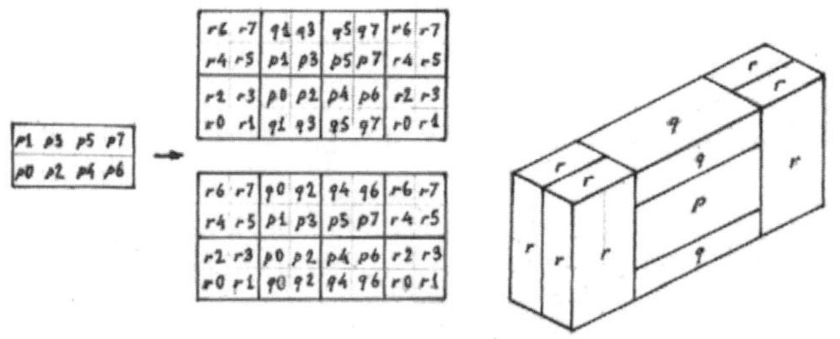

Figure 3. Labeling of protobricks.and unit cubes.

Fig. 4 shows generations 0 through 2 of the color code tiling starting with [00]; it reproduces brick 0 (yellow). Fig. 5 shows the first four layers (of eight) of generation 3 of the color code tiling starting with [00]; it clearly reproduces generation 1 of the inflation of brick 0 yellow, The remaining four layers repeat the first four ones. The tiling is rather repetitive as can be seen already from generation 2.

$$
\begin{array}{llllllll}
[00]\to\begin{matrix}\begin{bmatrix}32&33\\30&31\end{bmatrix}\\\begin{bmatrix}32&33\\30&31\end{bmatrix}\end{matrix} &
[01]\to\begin{matrix}\begin{bmatrix}36&37\\34&35\end{bmatrix}\\\begin{bmatrix}36&37\\34&35\end{bmatrix}\end{matrix} &
[02]\to\begin{matrix}\begin{bmatrix}00&02\\11&13\end{bmatrix}\\\begin{bmatrix}00&02\\10&12\end{bmatrix}\end{matrix} &
[03]\to\begin{matrix}\begin{bmatrix}11&13\\01&03\end{bmatrix}\\\begin{bmatrix}10&12\\01&03\end{bmatrix}\end{matrix} &
[04]\to\begin{matrix}\begin{bmatrix}04&06\\15&17\end{bmatrix}\\\begin{bmatrix}04&06\\14&16\end{bmatrix}\end{matrix} &
[05]\to\begin{matrix}\begin{bmatrix}15&17\\05&07\end{bmatrix}\\\begin{bmatrix}14&16\\05&07\end{bmatrix}\end{matrix} &
[06]\to\begin{matrix}\begin{bmatrix}32&33\\30&31\end{bmatrix}\\\begin{bmatrix}32&33\\30&31\end{bmatrix}\end{matrix} &
[07]\to\begin{matrix}\begin{bmatrix}36&37\\34&35\end{bmatrix}\\\begin{bmatrix}36&37\\34&35\end{bmatrix}\end{matrix} \\
[10]\to\begin{matrix}\begin{bmatrix}42&43\\42&43\end{bmatrix}\\\begin{bmatrix}40&41\\40&41\end{bmatrix}\end{matrix} &
[11]\to\begin{matrix}\begin{bmatrix}46&47\\46&47\end{bmatrix}\\\begin{bmatrix}44&45\\44&45\end{bmatrix}\end{matrix} &
[12]\to\begin{matrix}\begin{bmatrix}10&12\\10&12\end{bmatrix}\\\begin{bmatrix}00&12\\01&12\end{bmatrix}\end{matrix} &
[13]\to\begin{matrix}\begin{bmatrix}00&02\\01&03\end{bmatrix}\\\begin{bmatrix}11&13\\11&13\end{bmatrix}\end{matrix} &
[14]\to\begin{matrix}\begin{bmatrix}14&16\\14&16\end{bmatrix}\\\begin{bmatrix}04&06\\05&07\end{bmatrix}\end{matrix} &
[15]\to\begin{matrix}\begin{bmatrix}04&06\\05&07\end{bmatrix}\\\begin{bmatrix}15&17\\15&17\end{bmatrix}\end{matrix} &
[16]\to\begin{matrix}\begin{bmatrix}42&43\\42&43\end{bmatrix}\\\begin{bmatrix}40&41\\40&41\end{bmatrix}\end{matrix} &
[17]\to\begin{matrix}\begin{bmatrix}46&47\\46&47\end{bmatrix}\\\begin{bmatrix}44&45\\44&45\end{bmatrix}\end{matrix} \\
[20]\to\begin{matrix}\begin{bmatrix}53&53\\52&52\end{bmatrix}\\\begin{bmatrix}51&51\\50&50\end{bmatrix}\end{matrix} &
[21]\to\begin{matrix}\begin{bmatrix}57&57\\56&56\end{bmatrix}\\\begin{bmatrix}55&55\\54&54\end{bmatrix}\end{matrix} &
[22]\to\begin{matrix}\begin{bmatrix}22&22\\20&20\end{bmatrix}\\\begin{bmatrix}32&33\\30&31\end{bmatrix}\end{matrix} &
[23]\to\begin{matrix}\begin{bmatrix}32&33\\30&31\end{bmatrix}\\\begin{bmatrix}23&23\\21&21\end{bmatrix}\end{matrix} &
[24]\to\begin{matrix}\begin{bmatrix}26&26\\24&24\end{bmatrix}\\\begin{bmatrix}36&37\\34&35\end{bmatrix}\end{matrix} &
[25]\to\begin{matrix}\begin{bmatrix}36&37\\34&35\end{bmatrix}\\\begin{bmatrix}27&27\\25&25\end{bmatrix}\end{matrix} &
[26]\to\begin{matrix}\begin{bmatrix}53&53\\52&52\end{bmatrix}\\\begin{bmatrix}51&51\\50&50\end{bmatrix}\end{matrix} &
[27]\to\begin{matrix}\begin{bmatrix}57&57\\56&56\end{bmatrix}\\\begin{bmatrix}55&55\\54&54\end{bmatrix}\end{matrix} \\
[30]\to\begin{matrix}\begin{bmatrix}01&03\\00&02\end{bmatrix}\\\begin{bmatrix}01&03\\00&02\end{bmatrix}\end{matrix} &
[31]\to\begin{matrix}\begin{bmatrix}05&07\\04&06\end{bmatrix}\\\begin{bmatrix}05&07\\04&06\end{bmatrix}\end{matrix} &
[32]\to\begin{matrix}\begin{bmatrix}23&32\\21&30\end{bmatrix}\\\begin{bmatrix}22&32\\20&30\end{bmatrix}\end{matrix} &
[33]\to\begin{matrix}\begin{bmatrix}33&23\\31&21\end{bmatrix}\\\begin{bmatrix}33&22\\31&20\end{bmatrix}\end{matrix} &
[34]\to\begin{matrix}\begin{bmatrix}27&36\\25&34\end{bmatrix}\\\begin{bmatrix}26&36\\24&34\end{bmatrix}\end{matrix} &
[35]\to\begin{matrix}\begin{bmatrix}37&27\\35&25\end{bmatrix}\\\begin{bmatrix}37&26\\35&24\end{bmatrix}\end{matrix} &
[36]\to\begin{matrix}\begin{bmatrix}01&03\\00&02\end{bmatrix}\\\begin{bmatrix}01&03\\00&02\end{bmatrix}\end{matrix} &
[37]\to\begin{matrix}\begin{bmatrix}05&07\\04&06\end{bmatrix}\\\begin{bmatrix}05&07\\04&06\end{bmatrix}\end{matrix} \\
[40]\to\begin{matrix}\begin{bmatrix}11&13\\11&13\end{bmatrix}\\\begin{bmatrix}10&12\\10&12\end{bmatrix}\end{matrix} &
[41]\to\begin{matrix}\begin{bmatrix}15&17\\15&17\end{bmatrix}\\\begin{bmatrix}14&16\\14&16\end{bmatrix}\end{matrix} &
[42]\to\begin{matrix}\begin{bmatrix}53&42\\52&42\end{bmatrix}\\\begin{bmatrix}51&40\\50&40\end{bmatrix}\end{matrix} &
[43]\to\begin{matrix}\begin{bmatrix}43&53\\43&52\end{bmatrix}\\\begin{bmatrix}41&51\\41&50\end{bmatrix}\end{matrix} &
[44]\to\begin{matrix}\begin{bmatrix}57&46\\56&46\end{bmatrix}\\\begin{bmatrix}55&44\\54&44\end{bmatrix}\end{matrix} &
[45]\to\begin{matrix}\begin{bmatrix}47&57\\47&56\end{bmatrix}\\\begin{bmatrix}45&55\\45&54\end{bmatrix}\end{matrix} &
[46]\to\begin{matrix}\begin{bmatrix}11&13\\11&13\end{bmatrix}\\\begin{bmatrix}10&12\\10&12\end{bmatrix}\end{matrix} &
[47]\to\begin{matrix}\begin{bmatrix}15&17\\15&17\end{bmatrix}\\\begin{bmatrix}14&16\\14&16\end{bmatrix}\end{matrix} \\
[50]\to\begin{matrix}\begin{bmatrix}23&23\\21&21\end{bmatrix}\\\begin{bmatrix}22&22\\20&20\end{bmatrix}\end{matrix}, &
[51]\to\begin{matrix}\begin{bmatrix}27&27\\25&25\end{bmatrix}\\\begin{bmatrix}26&26\\24&25\end{bmatrix}\end{matrix} &
[52]\to\begin{matrix}\begin{bmatrix}52&52\\42&43\end{bmatrix}\\\begin{bmatrix}50&50\\40&41\end{bmatrix}\end{matrix} &
[53]\to\begin{matrix}\begin{bmatrix}42&43\\53&53\end{bmatrix}\\\begin{bmatrix}40&41\\51&51\end{bmatrix}\end{matrix} &
[54]\to\begin{matrix}\begin{bmatrix}56&56\\46&47\end{bmatrix}\\\begin{bmatrix}54&54\\44&45\end{bmatrix}\end{matrix} &
[55]\to\begin{matrix}\begin{bmatrix}46&47\\57&57\end{bmatrix}\\\begin{bmatrix}44&45\\55&55\end{bmatrix}\end{matrix} &
[56]\to\begin{matrix}\begin{bmatrix}23&23\\21&21\end{bmatrix}\\\begin{bmatrix}22&22\\20&20\end{bmatrix}\end{matrix} &
[57]\to\begin{matrix}\begin{bmatrix}27&27\\25&25\end{bmatrix}\\\begin{bmatrix}26&26\\24&25\end{bmatrix}\end{matrix}
\end{array}
\tag{2}
$$

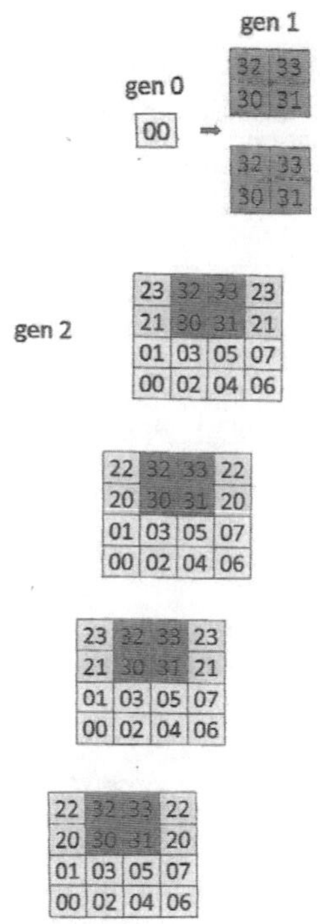

Figure 4. Generations 0 through 2 of the color code tiling reproducing brick 0 (yellow)..

IOP Conf. Series: Journal of Physics: Conf. Series **809** (2017) 012025 doi:10.1088/1742-6596/809/1/012025

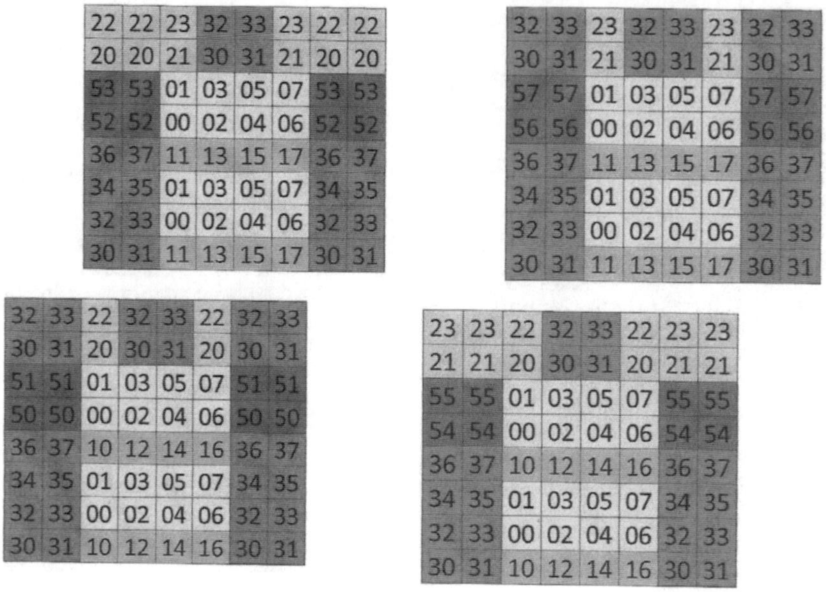

Figure 5. First four layers of generation 3 of the color code tiling reproducing the inflation of brick 0 (yellow)

.

References

[1] Ben-Abraham S I and Flom D 2016 Brick tiling *J. Phys. Conf. Series* (this volume)
[2] Robinson E A Jr 1999 On the table and the chair *Indag. Mathem. N.S.* **10** 581-599

Diffraction of a binary non-Pisot inflation tiling

Michael Baake[1,3] **and Uwe Grimm**[2,3]

[1] Fakultät für Mathematik, Universität Bielefeld, Postfach 100131, 33501 Bielefeld, Germany
[2] School of Mathematics and Statistics, The Open University, Milton Keynes MK7 6AA, UK
[3] School of Mathematics and Physics, University of Tasmania, Hobart TAS 7001, Australia

E-mail: `mbaake@math.uni-bielefeld.de`, `uwe.grimm@open.ac.uk`

Abstract. A one-parameter family of binary inflation rules in one dimension is considered. Apart from the first member, which is the well-known Fibonacci rule, no inflation factor is a unit. We identify all cases with pure point spectrum, and discuss the diffraction spectra of other members of the family. Apart from the trivial Bragg peaks at the origin, they have purely singular continuous diffraction.

Despite various open questions on details, the theory of substitutions with pure point spectrum is fairly well developed. In particular, for any given substitution on a finite alphabet, pure pointedness of the spectrum can be decided algorithmically; see [1] and references therein. For substitutions of constant length, the situation is even better due to Dekking's result [7] and its recent extension by Bartlett [6]. In general, however, the understanding of substitutions of non-constant length with mixed spectrum is still rudimentary. Recent work has indicated that progress in this direction is easier in the geometric setting of tiling spaces with natural tile sizes; see [3] and references therein. We adopt this point of view here, too.

In this contribution, we consider the family of primitive substitution rules on the binary alphabet $\{0,1\}$ defined by

$$\varrho_m : \quad 0 \mapsto 01^m \, , \; 1 \mapsto 0 \, , \quad \text{with } m \in \mathbb{N}.$$

The substitution matrix is $M_m = \left(\begin{smallmatrix} 1 & 1 \\ m & 0 \end{smallmatrix}\right)$ with eigenvalues $\lambda_m^{\pm} = \frac{1}{2}\bigl(1 \pm \sqrt{4m+1}\,\bigr)$, which are the roots of $\lambda^2 - \lambda - m = 0$. The frequency-normalised Perron–Frobenius (PF) eigenvector is $(1, \lambda_m^{+} - 1)^t / \lambda_m^{+}$, whose entries are the relative frequencies of the two letters. The corresponding left eigenvector reads $(\lambda_m^{+}, 1)$, which is our choice of the interval lengths for the corresponding geometric inflation rule. Up to scale, this is the unique choice to obtain a self-similar *inflation tiling* of the line from ϱ_m; see [4, Ch. 4] for background.

From now on, we will mainly work with the tiling system on the real line. Note that ϱ_1 defines the ubiquitous *Fibonacci tiling*, which is well-known to have pure point spectrum, both in diffraction and in dynamical sense [8, 4, 5]. For $m = 2$, we obtain a system that is equivalent to the *period doubling chain*, as can be seen by choosing $a \,\widehat{=}\, 0$ and $b \,\widehat{=}\, 11$ which establishes a mutual local derivability (MLD) rule; see [4, Secs. 4.5.1 and 9.4.4] for details on the period doubling system. More generally, λ_m^{+} is an integer if and only if $4m+1$ is a square. This precisely happens for $m = \ell(\ell+1)$ with $\ell \in \mathbb{N}$, giving $\lambda_m^{+} = \ell+1$ and $\lambda_m^{-} = -\ell$. Similar to the $m = 2$ case, any of these systems can be recoded as a constant length substitution, via $a \,\widehat{=}\, 0$ and $b \,\widehat{=}\, 1^{\ell+1}$. The induced substitution is $a \mapsto ab^{\ell}$, $b \mapsto a^{\ell+1}$, which has a coincidence in the first position.

Figure 1. Illustration of the self-similar inflation rule derived from ϱ_3.

Consequently, one has pure point spectrum by Dekking's criterion [7]. In all remaining cases, the PF eigenvalue is irrational, but fails to be a Pisot–Vijayaraghavan (PV) number, which means that none of the corresponding tilings can have non-trivial point spectrum [9, 2]. In particular, the only Bragg peak is the trivial one at $k = 0$. So far, we have the following result.

Theorem 1 *Consider the inflation tiling defined by* ϱ_m*. For* $m = 1$ *and* $m = \ell(\ell+1)$ *with* $\ell \in \mathbb{N}$*, the tiling has pure point diffraction, which can be calculated with the projection method. For all remaining cases, the pure point part of the diffraction consists of the trivial Bragg peak at* 0*, while the remainder of the diffraction is of continuous type.* □

The first example with continuous component emerges for $m = 3$, where $\lambda^{\pm} = \frac{1}{2}\big(1 \pm \sqrt{13}\,\big)$. Let us discuss this case in some more detail (the other non-PV cases are fairly analogous). The concrete inflation rule for this case is illustrated in Figure 1. With $\varrho := \varrho_3$, consider the bi-infinite fixed point w of ϱ^2 with legal seed $0|0$ obtained as

$$0|0 \overset{\varrho^2}{\longmapsto} w^{(1)} = 0111000|0111000 \overset{\varrho^2}{\longmapsto} \ldots \overset{\varrho^2}{\longmapsto} w^{(i)} \xrightarrow{\; i\to\infty \;} w = \varrho^2(w);$$

see [8, 4] for background. The corresponding tiling, via the left endpoints of the two types of intervals, leads to a Delone set

$$\varLambda^{w} = \{\ldots, -1-3\lambda, -3\lambda, -2\lambda, -\lambda, 0, \lambda, 1+\lambda, 2+\lambda, 3+\lambda, 3+2\lambda, \ldots\} \subset \mathbb{Z}[\lambda],$$

where $\lambda = \lambda_3^{+}$ from now on. Defining the hull $\mathbb{Y} := \overline{\{t + \varLambda^{w} \mid t \in \mathbb{R}\}}^{\mathrm{LT}}$ with the closure being taken in the local topology (LT), we obtain a topological dynamical system $(\mathbb{Y}, \mathbb{R})$ under the translation action of $\mathbb{R}$. This system is *strictly ergodic*, which means it is *minimal* (because $\mathbb{Y}$ coincides with the local indistinguishability (LI) class of $\varLambda^{w}$) and *uniquely ergodic* (since there is a unique probability measure $\mu_{\mathbb{Y}}$, defined by the uniformly existing patch frequencies).

One important consequence is that every $\varLambda \in \mathbb{Y}$ has the same autocorrelation and the same diffraction measure as $\varLambda^{w}$, which are thus also called the autocorrelation and diffraction of the dynamical system. More generally, given any $\varLambda \in \mathbb{Y}$, we consider the weighted Dirac comb $\omega := \sum_{x\in\varLambda} u(x)\delta_x$ with general complex weights $u(x) \in \{u_0, u_1\}$ according to the interval type. Then, the corresponding autocorrelation γ_u is of the form

$$\gamma_u = \sum_{z\in\varLambda-\varLambda} \eta_u(z)\,\delta_z \quad \text{with} \quad \eta_u(z) = \lim_{r\to\infty} \frac{1}{2r} \sum_{y,y+z\in\varLambda_r} \overline{u(y)}\,u(y+z),$$

with $\varLambda_r := \varLambda \cap [-r, r]$.

Since we do not have any projection method at our disposal for the further analysis, other tools are needed. It has recently been shown in [3] that the *pair correlation functions* of a primitive inflation tiling satisfy a set of *exact renormalisation relations* that help to unravel the spectral type of the diffraction. Partitioning $\varLambda = \varLambda^{(0)} \dot\cup \varLambda^{(1)}$ into the two point types, one can define the pair correlation functions as

$$\nu_{ij}(z) = \lim_{r\to\infty} \frac{\mathrm{card}\big(\varLambda_r^{(i)} \cap (\varLambda_r^{(j)} - z)\big)}{\mathrm{card}(\varLambda_r)} = \frac{1}{\mathrm{dens}(\varLambda)} \lim_{r\to\infty} \frac{\mathrm{card}\big(\varLambda_r^{(i)} \cap (\varLambda_r^{(j)} - z)\big)}{2r}$$

for $i, j \in \{0, 1\}$. These functions are well-defined for any $z \in \mathbb{R}$, non-negative and satisfy the symmetry relations $\nu_{ij}(z) = \nu_{ji}(-z)$. Moreover, $\nu_{ij}(z) > 0$ if and only if $z \in \varLambda^{(j)} - \varLambda^{(i)}$, which is a consequence of strict ergodicity.

The autocorrelation coefficients $\eta_u(z)$ can now be expressed as

$$\eta_u(z) \; = \; \mathrm{dens}(\varLambda) \sum_{i,j \in \{0,1\}} \overline{u_i}\, \nu_{ij}(z)\, u_j$$

and similar expressions will later emerge for the diffraction measure $\widehat{\gamma_u}$. As follows from [3, 2], one has the following result.

Theorem 2 *The pair correlations functions ν_{ij} satisfy the linear renormalisation equations*

$$\begin{aligned}
\nu_{00}(z) &= \frac{1}{\lambda}\Big(\nu_{00}\big(\tfrac{z}{\lambda}\big) + \nu_{01}\big(\tfrac{z}{\lambda}\big) + \nu_{10}\big(\tfrac{z}{\lambda}\big) + \nu_{11}\big(\tfrac{z}{\lambda}\big)\Big),\\
\nu_{01}(z) &= \frac{1}{\lambda}\Big(\nu_{00}\big(\tfrac{z-\lambda}{\lambda}\big) + \nu_{00}\big(\tfrac{z-1-\lambda}{\lambda}\big) + \nu_{00}\big(\tfrac{z-2-\lambda}{\lambda}\big) + \nu_{10}\big(\tfrac{z-\lambda}{\lambda}\big) + \nu_{10}\big(\tfrac{z-1-\lambda}{\lambda}\big) + \nu_{10}\big(\tfrac{z-2-\lambda}{\lambda}\big)\Big),\\
\nu_{10}(z) &= \frac{1}{\lambda}\Big(\nu_{00}\big(\tfrac{z+\lambda}{\lambda}\big) + \nu_{00}\big(\tfrac{z+1+\lambda}{\lambda}\big) + \nu_{00}\big(\tfrac{z+2+\lambda}{\lambda}\big) + \nu_{01}\big(\tfrac{z+\lambda}{\lambda}\big) + \nu_{01}\big(\tfrac{z+1+\lambda}{\lambda}\big) + \nu_{01}\big(\tfrac{z+2+\lambda}{\lambda}\big)\Big),\\
\nu_{11}(z) &= \frac{1}{\lambda}\Big(3\,\nu_{00}\big(\tfrac{z}{\lambda}\big) + 2\,\nu_{00}\big(\tfrac{z+1}{\lambda}\big) + 2\,\nu_{00}\big(\tfrac{z-1}{\lambda}\big) + \nu_{00}\big(\tfrac{z+2}{\lambda}\big) + \nu_{00}\big(\tfrac{z-2}{\lambda}\big)\Big).
\end{aligned}$$

Subject to the condition that the support of each ν_{ij} is $\varLambda^{(j)} - \varLambda^{(i)}$, the solution space of this infinite system of linear equations is one-dimensional. □

Via $\varUpsilon_{ij} := \sum_{z \in S_{ij}} \nu_{ij}(z)\, \delta_z$, the pair correlation functions are turned into pure point measures. The autocorrelation measure γ_u can now be written as $\gamma_u(\mathcal{E}) = \mathrm{dens}(\varLambda) \sum_{i,j \in \{0,1\}} \overline{u_i}\, \varUpsilon_{ij}(\mathcal{E})\, u_j$, where $\mathcal{E} \subset \mathbb{R}$ is any bounded Borel set. Taking the Fourier transform, one obtains the diffraction measure as

$$\widehat{\gamma_u}(\mathcal{E}) \; = \; \mathrm{dens}(\varLambda) \sum_{i,j \in \{0,1\}} \overline{u_i}\, \widehat{\varUpsilon}_{ij}(\mathcal{E})\, u_j,$$

where the Fourier transform of each term can be shown to exist [3]. Now, the renormalisation relations from Theorem 2 induce measure-valued relations for the $\widehat{\varUpsilon}_{ij}$, which have to be satisfied for each part of their Lebesgue decompositions separately. An analysis of the asymptotic behaviour of the absolutely continuous components shows that the only contribution compatible with local integrability of the Radon–Nikodym densities and the translation boundedness of $\widehat{\gamma_u}$ is the trivial one, which means that no absolutely continuous component is possible [2]. Consequently, one has the following result.

Theorem 3 *The diffraction measure $\widehat{\gamma_u}$, which is the same for all $\varLambda \in \mathbb{Y}$, has the pure point part $\big|\frac{2\lambda-1}{13} u_0 + \frac{7-\lambda}{13} u_1\big|^2\, \delta_0$. The remainder of $\widehat{\gamma_u}$ is singular continuous.* □

To give an impression of the singular continuous part, let us choose $u_0 = 1 - \lambda$ and $u_1 = 1$. With this choice, the Bragg peak at 0 is extinct, so $\widehat{\gamma_u}$ is purely singular continuous, with $\eta_u(0) = (6\lambda - 3)/13 \approx 0.832$. Consequently, the *distribution function* F defined by $F(x) := \widehat{\gamma_u}\big([0, x]\big)$ is continuous, with average slope given by $\eta_u(0)$; see Figure 2 for an illustration. Note that F is strictly increasing, despite the appearance of 'flat' regions which resemble plateaux.

Let us close by commenting on the other non-Pisot members of our inflation tiling family. So, let m be any integer such that the pure point part of the diffraction, according to Theorem 1,

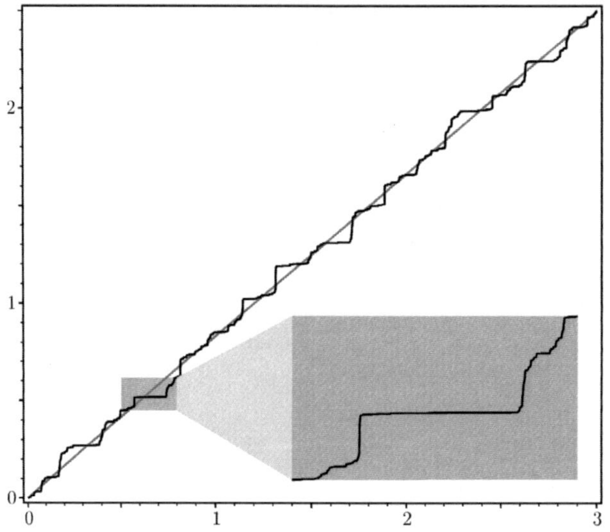

Figure 2. Sketch of the distribution function F for the singular continuous diffraction measure discussed in the text. The average slope of F is $\eta_u(0)$, which is indicated by a straight line.

is trivial. Then, with our choice of interval lengths from the beginning, the Bragg peak at the origin has intensity

$$I_0 = \left|\operatorname{dens}(\varLambda)\,(u \cdot v_{\mathrm{PF}})\right|^2 = \frac{\left|u_0 + (\lambda_m^+ - 1)u_1\right|^2}{4m+1},$$

where $\operatorname{dens}(\varLambda) = \frac{\lambda_m^+}{2\lambda_m^+ - 1} = \frac{\lambda_m^+ + 2m}{4m+1}$ and v_{PF} is the PF frequency vector from above.

For each such inflation tiling, the pair correlation functions are again well-defined, and satisfy a set of exact linear renormalisation relations in analogy to Theorem 2. Completing the corresponding analysis on the Fourier side, our analysis indicates that we can never have an absolutely continuous component. A numerical calculation of the distribution function analogous to F above produces graphs that are very similar to the one shown in Figure 2.

References

[1] Akiyama S, Gähler F and Lee J-Y 2014 Determining pure discrete spectrum for some self-affine tilings *Discr. Math. Theor. Comput. Sci.* **16** 305–316; `arXiv:1403.0362`

[2] Baake M, Frank NP, Grimm U and Robinson EA 2016 Geometric properties of a non-Pisot inflation point set and absence of absolutely continuous diffraction *Preprint*

[3] Baake M and Gähler F 2016 Pair correlations of aperiodic inflation rules via renormalisation: Some interesting examples *Top. Appl.* **205** 4–27; `arXiv:1511.00885`

[4] Baake M and Grimm U 2013 *Aperiodic Order. Volume 1: A Mathematical Invitation* (Cambridge University Press, Cambridge)

[5] Baake M and Lenz D 2016 Spectral notions of aperiodic order *Discr. Cont. Dynam. Syst. S*, in press; `arXiv:1601.06629`

[6] Bartlett A 2014 Spectral theory of $\mathbb{Z}^d$ substitutions *Ergod. Th. & Dynam. Syst.*, in press; `arXiv:1410.8106`

[7] Dekking FM 1978 The spectrum of dynamical systems arising from substitutions of constant length *Z. Wahrscheinlichkeitsth. verw. Geb.* **41** 221–239

[8] Queffélec M 2010 *Substitution Dynamical Systems — Spectral Analysis* LNM 1294, 2nd ed. (Springer, Berlin)

[9] Solomyak B 1997 Dynamics of self-similar tilings, *Ergod. Th. & Dynam. Syst.* **17** 695–738 and **19** (1999) 1685 (Erratum)

IOP Conf. Series: Journal of Physics: Conf. Series **809** (2017) 012027 doi:10.1088/1742-6596/809/1/012027

Substitution-based sequences with absolutely continuous diffraction

Lax Chan and Uwe Grimm
School of Mathematics and Statistics, The Open University, Milton Keynes MK7 6AA, UK
E-mail: lax.chan@open.ac.uk, uwe.grimm@open.ac.uk

Abstract. Modifying Rudin's original construction of the Rudin–Shapiro sequence, we derive a new substitution-based sequence with purely absolutely continuous diffraction spectrum.

1. Introduction

Substitution dynamical systems are widely used as toy models for aperiodic order in one dimension [1]. The binary Rudin–Shapiro sequence [2, 3, 4] is a paradigm for a substitution-based structure with (in its balanced weight case) purely absolutely continuous diffraction spectrum; we refer to [5] for details and background. Indeed, this deterministic sequence has the stronger property that its two-point correlations vanish for *any* non-zero distance, as it would be the case for a random sequence [6]. While currently only very few examples of substitution-based sequences with this property are known, a systematic generalisation in terms of Hadamard matrices [7] allows the construction of one-dimensional as well as higher-dimensional examples.

We start by reviewing Rudin's original construction [4], and modify it to obtain a new example of a substitution-based structure with absolutely continuous diffraction spectrum. Our approach differs from the construction in [7]. We then use recent work of Bartlett [5] to investigate the properties of the new system in more detail.

2. Original construction of the Rudin–Shapiro sequence

In 1958, Salem [4] asked the following question: is it possible to find a sequence $(\varepsilon_n)_{n\in\mathbb{N}} \in \{\pm 1\}^{\mathbb{N}}$ such that there is a constant $C > 0$ for which

$$\sup_{\theta\in\mathbb{R}} \left| \sum_{n<N} \varepsilon_n \, e^{2\pi i n \theta} \right| \leq C\sqrt{N} \tag{1}$$

holds for any positive integer N? This is known as the 'root N' property, which implies absolutely continuous diffraction of the sequence $(\varepsilon_n)_{n\in\mathbb{N}}$ by an application of [8, Prop. 4.9]. Shapiro [2] and Rudin [4], and indeed earlier Golay [3], gave positive answers to the question by constructing what is now known as the Rudin–Shapiro sequence. We briefly review Rudin's construction to obtain the substitution dynamical system and its balanced weight version (with weights in ± 1).

We start by defining polynomials $P_k(x)$ and $Q_k(x)$ of degree 2^k for $k \in \mathbb{N}_0$ recursively by

$$\begin{aligned} P_{k+1}(x) &= P_k(x) + x^{2^k} Q_k(x), \\ Q_{k+1}(x) &= P_k(x) - x^{2^k} Q_k(x), \end{aligned} \tag{2}$$

with $P_0(x) = Q_0(x) = x$. Note that from Eq. (2) it is clear that the first 2^k terms of $P_{k+1}(x)$ and of $Q_{k+1}(x)$ coincide with those of $P_k(x)$, and the remaining terms differ by a sign. By construction, $P_k(x)$ is of the form

$$P_k(x) = \sum_{n=1}^{2^k} \varepsilon_n \, x^n, \tag{3}$$

so we can define a binary sequence $(\varepsilon_n)_{n\in\mathbb{N}} \in \{\pm 1\}^{\mathbb{N}}$ from the corresponding coefficients. This is the binary Rudin–Shapiro sequence. For example, $P_3(x) = x+x^2+x^3-x^4+x^4(x+x^2-x^3+x^4)$, from which we read off the sequence $111\bar{1}11\bar{1}1$ with $\bar{1} = -1$. The main ingredient in the proof of property (1) for this sequence is the parallelogram law for $\alpha, \beta \in \mathbb{C}$,

$$|\alpha+\beta|^2 + |\alpha-\beta|^2 = 2|\alpha|^2 + 2|\beta|^2,$$

applied to $P_k(e^{2\pi i\theta})$; see [4] for details.

Often, the Rudin–Shapiro sequence is defined by a four-letter substitution rule and a subsequent reduction map from four to two letters. The underlying four-letter substitution on the alphabet $\{A, B, C, D\}$ can be read off from the recursion (2), noting that the recursion implies the concatenation of the sequences corresponding to P_k and Q_k. Associating letters A and B to the coefficients in P and Q and the letters C and D to those of $-Q$ and $-P$, respectively, this gives rise to the four-letter substitution rule

$$\varrho\colon \quad A \mapsto AB, \quad B \mapsto AC, \quad C \mapsto DB, \quad D \mapsto DC,$$

which corresponds to the standard four-letter Rudin–Shapiro subsitution. Iterating the sequence on the initial seed A gives

$$A \mapsto AB \mapsto ABAC \mapsto ABACABDB \mapsto \ldots$$

which, under the mapping $A \mapsto 1$, $B \mapsto 1$, $C \mapsto -1$, $D \mapsto -1$ (which corresponds to the choice of signs above) produces the binary Rudin–Shapiro sequence $(\varepsilon_n)_{n\in\mathbb{N}}$.

3. Generalising Rudin's construction

Let us now modify Rudin's original argument by considering the following system

$$\begin{aligned} P_{k+1}(x) &= P_k(x) + (-1)^k x^{2^k} Q_k(x), \\ Q_{k+1}(x) &= P_k(x) - (-1)^k x^{2^k} Q_k(x), \end{aligned} \tag{4}$$

starting again from $P_0(x) = Q_0(x) = x$. Since from now on we shall only look at this set of recursion relations, we use the same notation as above. What has changed in comparison with Eq. (2) is that we swap the sign in the recursion relation in each step. Since the sign does not affect the argument used in Rudin's proof [4], it is straightforward to show that the parallelogram law can again be used to prove that the corresponding *new* sequence of coefficients $(\varepsilon_n)_{n\in\mathbb{N}}$, defined as above via Eq. (3), also satisfies the root N property.

Can we read off the corresponding substitution rule as for the Rudin–Shapiro case? Indeed this is possible, but it is slightly more complicated, because the recursion relations in Eq. (4) alternate. One way to deal with that is to look at two consecutive steps together,

$$\begin{aligned} P_{k+2}(x) &= P_k(x) + (-1)^k x^{2^k} Q_k(x) + (-1)^{k+1} x^{2\cdot 2^k} P_k(x) + x^{3\cdot 2^k} Q_k(x), \\ Q_{k+2}(x) &= P_k(x) + (-1)^k x^{2^k} Q_k(x) - (-1)^{k+1} x^{2\cdot 2^k} P_k(x) - x^{3\cdot 2^k} Q_k(x). \end{aligned} \tag{5}$$

Choosing k to be even, say, and associating again four letters A, B, C, D to the sequences corresponding to $P, Q, -Q, -P$, we can read off the substitution

$$\sigma\colon \quad A \mapsto ABDB, \quad B \mapsto ABAC, \quad C \mapsto DCDB, \quad D \mapsto DCAC, \tag{6}$$

which is now a constant length substitution of length four, because we used a double step of the recursion. Therefore Eq. (5) corresponds to concatenation of four sets of coefficients. As before, the binary sequence is obtained from iterating the substitution on the initial letter A;

$$A \mapsto ABDB \mapsto ABDBABACDCACABAC \mapsto \cdots . \tag{7}$$

By mapping A and B to 1 and mapping C and D to $\bar{1} = -1$, we obtain $11\bar{1}1111\bar{1}\bar{1}\bar{1}1\bar{1}111\bar{1}\ldots$ as the initial part of our new binary sequence.

Clearly, we can generate infinitely many such examples by changing the signs in the original recursion relation (2) in more complicated ways. Any finite sequence of signs, when used periodically, will give rise to a substitution-based system. However, the length of the substitution will increase with the length of the sign sequence. There neither appears to be an obvious relation between the different Rudin–Shapiro type sequences obtained from this construction, nor between these sequences and those derived from Frank's construction [7]. In the latter case, the number of letters in the alphabet increases with the size of the Hadamard matrix, whereas all our sequences will only use four letters. Having said that, it would be possible to use more letters rather than consider multiple recursion steps, so the relation between these two approaches still needs to be analysed in more detail.

4. Spectral properties

Since we have Rudin's original argument at our disposal, we know that this new binary sequence satisfies the root N property (1), and hence has (in the balanced weight case with weights ± 1) purely absolutely continuous diffraction. Nevertheless, we would like to use the remainder of this paper to generalise this result by applying Bartlett's algorithm [9] to our new (four-letter substitution) sequence. This allows us to verify that the four-letter substitution sequence (in the balanced weight case) also has absolutely continuous diffraction spectrum only, whereas Rudin's argument only covers the binary case. Due to space constraints, we cannot introduce all quantities here, and instead refer to [9] for definitions and further details.

Theorem 1. *The balanced weight sequence derived from the substitution rule σ of Eq. (6) has purely absolutely continuous diffraction spectrum.*

Proof. The four instruction matrices R_i and the substitution matrix M can be read off of the substitution rule (6). They are given by

$$R_0 = \begin{pmatrix} 1&1&0&0\\ 0&0&0&0\\ 0&0&0&0\\ 0&0&1&1 \end{pmatrix}, \quad R_1 = \begin{pmatrix} 0&0&0&0\\ 1&1&0&0\\ 0&0&1&1\\ 0&0&0&0 \end{pmatrix}, \quad R_2 = \begin{pmatrix} 0&1&0&1\\ 0&0&0&0\\ 0&0&0&0\\ 1&0&1&0 \end{pmatrix}, \quad R_3 = \begin{pmatrix} 0&0&0&0\\ 1&0&1&0\\ 0&1&0&1\\ 0&0&0&0 \end{pmatrix},$$

with $M = R_0 + R_1 + R_2 + R_3$. As M^2 has positive entries only, the substitution is primitive. The second iterate of the seed A obtained in Eq. (7) shows that the letter A can be preceded by either B or C. Hence A has two distinct neighbourhoods and, by Pansiot's lemma [9, Lem. 3.6], the substitution is aperiodic.

Since we are dealing with a constant-length substitution, the leading (Perron–Frobenius) eigenvalue [5, Thm. 2.2] of M is $\lambda_{\mathrm{PF}} = 4$ with (statistically normalised) eigenvector and $u = \frac{1}{4}(1,1,1,1)$. By applying [9, Thm. 4.3], we obtain $\widehat{\Sigma}(0) = \frac{1}{4}\sum_{\alpha} e_{\alpha\alpha}$, where $e_{\alpha\alpha}$ denotes the unit vector corresponding to the word $\alpha\alpha$. As σ is a length four substitution, we have $\Delta_1(1) = \{3\}$. Using [9, Thm. 4.3], we obtain

$$\begin{aligned}\widehat{\Sigma}(1) &= \tfrac{1}{8}(0,1,1,0,1,0,0,1,1,0,0,1,0,1,1,0) = \widehat{\Sigma}(3),\\ \widehat{\Sigma}(2) &= \tfrac{1}{8}(1,0,0,1,0,1,1,0,0,1,1,0,1,0,0,1) = \widehat{\Sigma}(4).\end{aligned}$$

One can then verify the following recursive relations

$$\widehat{\Sigma}(4n) = \widehat{\Sigma}(4), \quad \widehat{\Sigma}(4n+1) = \widehat{\Sigma}(1), \quad \widehat{\Sigma}(4n+2) = \widehat{\Sigma}(2), \quad \widehat{\Sigma}(4n+3) = \widehat{\Sigma}(3). \tag{8}$$

Using [9, Prop. 2.2], we calculate the ergodic decomposition of the bi-substitution $\sigma \otimes \sigma$. We obtain $E_1 = \{AA, BB, CC, DD\}$ and $E_2 = \{AD, DA, BC, CB\}$ as the two ergodic classes and $T = \{AB, AC, BA, BD, CA, CD, DB, DC\}$ as the transient part. From [9, Lem. 4.7], we get

$$v = w_1 \begin{pmatrix} 1 & 0 & 0 & 0 \\ 0 & 1 & 0 & 0 \\ 0 & 0 & 1 & 0 \\ 0 & 0 & 0 & 1 \end{pmatrix} + w_2 \begin{pmatrix} 0 & 0 & 0 & 1 \\ 0 & 0 & 1 & 0 \\ 0 & 1 & 0 & 0 \\ 1 & 0 & 0 & 0 \end{pmatrix} + \frac{w_1 + w_2}{2} \begin{pmatrix} 0 & 1 & 1 & 0 \\ 1 & 0 & 0 & 1 \\ 1 & 0 & 0 & 1 \\ 0 & 1 & 1 & 0 \end{pmatrix}.$$

Diagonalising the matrix v, we obtain

$$v_d = \begin{pmatrix} 2(w_1 + w_2) & 0 & 0 & 0 \\ 0 & w_1 - w_2 & 0 & 0 \\ 0 & 0 & w_1 - w_2 & 0 \\ 0 & 0 & 0 & 0 \end{pmatrix}.$$

Setting $w_1 = 1$, strong semi-positivity is equivalent to w_2 satisfying $-1 \leq w_2 \leq 1$. The extreme points are then given by $(w_1, w_2) = (1, 1)$ and $(w_1, w_2) = (1, -1)$. Thus the extreme rays are

$$\begin{aligned} v_1 &= (1,1,1,1,1,1,1,1,1,1,1,1,1,1,1,1),\\ v_2 &= (1,0,0,-1,0,1,-1,0,0,-1,1,0,-1,0,0,1).\end{aligned}$$

As usual, $\lambda_{v_1} = \delta_0$, which gives rise to the pure point component of the spectrum. Using Eq. (8), one checks that $\widehat{\lambda_{v_2}}(k) = 0$ for all $k \neq 0$, which gives us the absolutely continuous component. Thus, we have a purely absolute continuous spectrum in the balanced weight case, in which the pure point component is extinguished. □

Acknowledgments

The authors would like to thank Ian Short for many helpful comments. L.C. would like to thank the organising committee for a young scientist award to attend ICQ13 in Kathmandu.

References

[1] Pytheas Fogg N 2002 *Substitutions in Dynamics, Arithmetics and Combinatorics* (Springer, Berlin)
[2] Shapiro H 1951 *Extremal Problems for Polynomials and Power Series* Masters thesis (MIT, Boston)
[3] Golay MJE 1951 Statistic multislit spectrometry and its application to the panoramic display of infrared spectra *J. Optical Soc. America* **41** 468–472
[4] Rudin W 1959 Some theorems on Fourier coefficients *Proc. Amer. Math. Soc.* **10** 855–859
[5] Baake M and Grimm U 2013 *Aperiodic Order. Vol. 1. A Mathematical Invitation* (Cambridge University Press, Cambridge)
[6] Baake M and Grimm U 2009 Kinematic diffraction is insufficient to distinguish order from disorder *Phys. Rev. B* **79** 020203 and **80** 029903 (Erratum)
[7] Frank N 2003 Substitution sequences in $\mathbb{Z}^d$ with a non-simple Lebesgue component in the spectrum *Ergod. Th. & Dynam. Syst.* **23** 519–532
[8] Queffélec M 2010 *Substitution Dynamical Systems — Spectral Analysis* 2nd ed (Springer, Berlin)
[9] Bartlett A 2014 Spectral theory of $\mathbb{Z}^d$ substitutions *Ergod. Th. & Dynam. Syst.* (to appear, preprint arXiv:1410.8106)

An ancient rule for constructing dodecagonal quasiperiodic patterns

Rima Ajlouni[1]

[1]School of Architecture, University of Utah, Salt Lake City, UT 84112, USA

E-mail: ajlouni@arch.utah.edu

Abstract. The discovery of complex dodecagonal patterns in historical Islamic architecture is generating a renewed interest into understanding the mathematical principles of traditional Islamic geometry. By employing a compass and a straightedge, ancient craftsmen utilized consistent design principles that allowed for diverse geometric expressions to be realized throughout the ancient world. Derived from these principles, a global multi-level structural model is proposed that provides a general guiding principle for constructing a wide variety of infinite dodecagon-based quasiperiodic patterns.

1. Introduction

The discovery of complex patterns in ancient Islamic architecture with symmetries that are incompatible with periodicity is providing a unique window into understanding the traditional principles of ancient geometry. To date three types of such symmetries were documented. These include; 10-fold [1-4], 8-fold [5-6] and 12-fold [7]. While many examples of decagonal and octagonal patterns were identified, only one instance of a dodecagonal cartwheel has been found. In 2011 Makovicky and Makovicky recognized the first ancient dodecagonal pattern in the Zaouïa of Moulay Idriss II in Fez, which they analyzed based on a type of Ammann quasilattice [7]. It is worth to note that there is no evidence to suggest that ancient craftsmen were aware of the concept of infinite tilings with long-range quasiperiodic order. Today, much is unknown about the traditional creative process by which these complicated patterns were conceived or applied. The question of which method is the best suited to explain the rich diversity of such patterns in Islamic Architecture remains at the heart of an ongoing debate [8]. While considerable effort has been invested into analyzing these designs based on modern mathematical theories (tiling, matching, grid, substitution, etc.), only few attempts have been made to bring forward a comprehensive hypotheses for explaining the traditional design process based on the nature of the design tools themselves; the compass and the straightedge. Traditionally, the manipulation of circles, straight lines and their intersections provides the underlying structure for guiding the creative development of the different designs. Derived from these simple principles, this paper proposes a global multi-level hierarchical structural model that is able to explain the underlying logic of traditional patterns in ancient Islamic architecture, as well as to provide a general guiding method for constructing a wide variety of infinite dodecagonal quasiperiodic patterns.

2. An ancient dodecagonal quasiperiodic pattern

The only known example of a complex dodecagonal pattern in Islamic architecture was found in the tympanum of the entrance of the Zaouïa Moulay Idriss II in Fez [7]. The central cartwheel portion (figure

1a) exhibits 12-fold rotational symmetry. Based on my examination of a wide range of complex patterns (5-fold, 10-fold and 8-fold) [4, 6, 9], a multi-level structural model is used to construct the dodecagonal quasiperiodic order. Governed by a tightly controlled proportional system, dodecagonal patterns can be generated based on a combination of two basic design elements; an underlying 'behind the scenes' structural network and the repeating 'seed units'. In this creative system, the network is responsible for defining the type of symmetry by providing the underlying structure for the spatial distribution of the 'seed units', while the internal formations of the 'seed units' derive the different design variations. Constructing the global structure of a dodecagonal quasiperiodic pattern requires building a progression of the structural network in multiple hierarchies, in which every hierarchy is built on the previous one.

To demonstrate this process, Figure 1 shows the construction process of the first hierarchy based on the dodecagonal pattern at the entrance of the Zaouïa Moulay Idriss II in Fez. Using a compass and a straight edge, a framework of nested dodecagrams is constructed (figure 1b). The relationship between each level in this network and the corresponding dodecagon, servers a critical role in maintaining a relational aspect ratio between all parts of the system. In this system, the size of the smallest dodecagon determines the size of the repeating 'seed polygons', allowing them to fit precisely in place without creating gaps or overlapping. Figure 1c shows the structural network after adapting the central portion to match the internal design of the ancient pattern. By mapping the central point of the 'seed polygons' to the intersection points of the structural network, a distribution map of the seed polygons is created (figure 1c). It is important to note that adjustments to some edge locations of the 'seed polygons' are made to match the ancient pattern (figure 1d). Figure 1e shows the process of constructing the internal design of the 'seed polygons' and the connecting formations. Such process follows a design logic that is similar to the process used for constructing the underlying structural network. As mentioned earlier, the internal design of the 'seed polygons' and its connecting units can assume different formations without affecting the dodecagonal symmetry. Such flexibility allows for a wide spectrum of designs to emerge by either manipulating the construction of the structural network or the internal design of the 'seed polygons'. The final formations of the first hierarchy of the dodecagonal pattern is shown in figures 1f and 1g. Figure 1h shows two alternative design options for the internal portion of the pattern. Figure 1i shows the dodecagonal pattern using the structural network in figures 1b and one of the alternative design options in figure 1h.

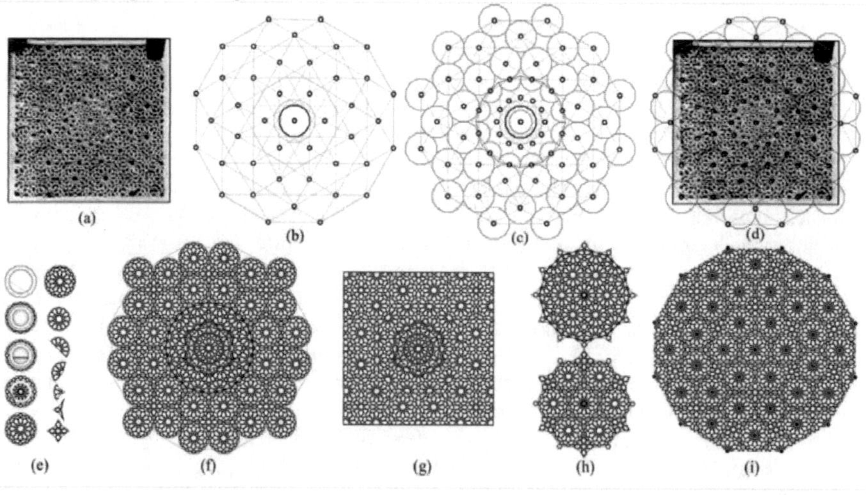

Figure 1. The construction process of the first hierarchy based on the dodecagonal pattern at the entrance of the Zaouïa Moulay Idriss II in Fez. The image used in (a) and (d) is taken from Makovicky and Makovicky [7] "Reproduced with permission of the International Union of Crystallography".

To construct the second hierarchy of the dodecagonal pattern, a new generation of the structural network is constructed (figure 2a). In this level, the final constructed pattern of the first hierarchy (figure 1i) acts as a 'seed unit' for the second hierarchy. The design of the connecting units are flexible and can take different formations including being part of the main 'seed unit' (figure 2b). The final dodecagonal pattern of the second hierarchy is shown in figure 2c. Constructing the infinite global empire of the dodecagonal quasiperiodic pattern requires building an infinite progression of hierarchies, in which, each new level is built on the previous one. This way a dodecagonal quasiperiodic pattern can grow ad infinitum without any gaps or imperfections.

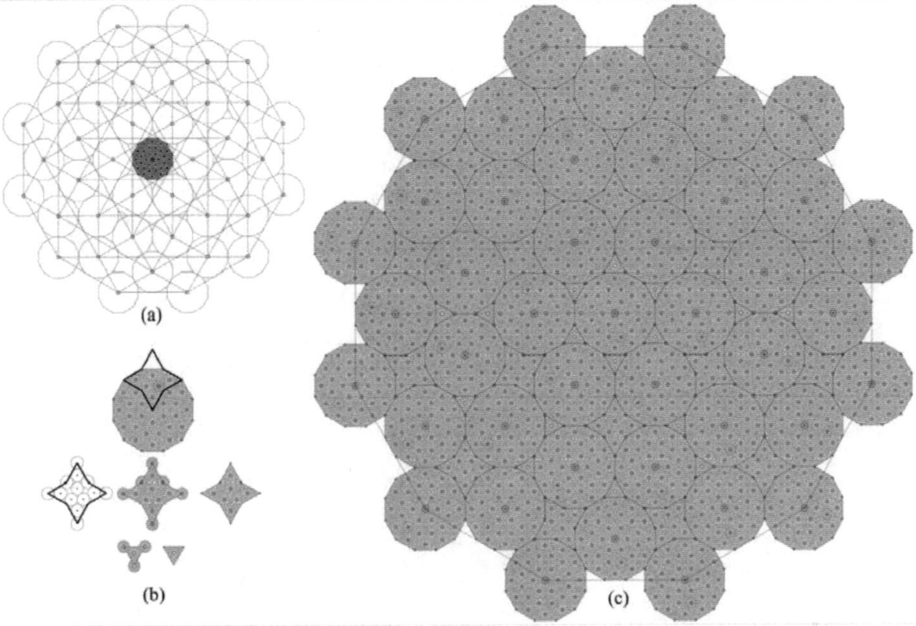

Figure 2. The construction process of the second hierarchy of the dodecagonal quasiperiodic pattern.

3. Constructing new dodecagonal quasiperiodic patterns

This section demonstrates the process of using the multi-level hierarchical method for constructing new variations of dodecagonal quasiperiodic patterns. The process of constructing the first hierarchy of the dodecagonal pattern starts by building a progression of nested dodecagrams. One alternative network design is shown in figure 3a. This network, which can be easily constructed by a using a compass and a straightedge, provides the underlying structural grid for mapping the locations of the 'seed polygons'. In this tightly proportional system, the size of the 'seed polygons' is defined by the generated central dodecagon of the network. This specific size ratio allows for the edges of the seed polygons to fit precisely in place without overlapping. By mapping the locations of the 'seed polygons' to specific intersection points of the structural network, a distribution map of the generalized 'seed polygons' is created (Figure 3b). In addition to the main 'seed polygons', connecting units are needed to fill-in the remaining gaps between the 'seed polygons'. In this specific case, three different connecting units are formed (figure 3c). It is worth noting that by connecting this specific formation of intersection points, a tessellation of three different tiles is created: a square, a thick rhomb and a thin rhomb (figure 3d). While the internal design of the 'seed polygons' can take different forms, using the same three-tile combinations for the internal design of the 'seed polygon' makes it possible to generate dodecagonal patterns that are self-similar (figure 3e). The final dodecagonal pattern of the first hierarchy is shown in figure 3f. Figure 3g shows some alternative design variations of the 'seed polygons'. By manipulating the internal design of the 'seed polygons' and its connecting formations, this flexible method allows for a wide spectrum of pattern variations while maintaining the symmetry of the underlying structural network (figure 3h).

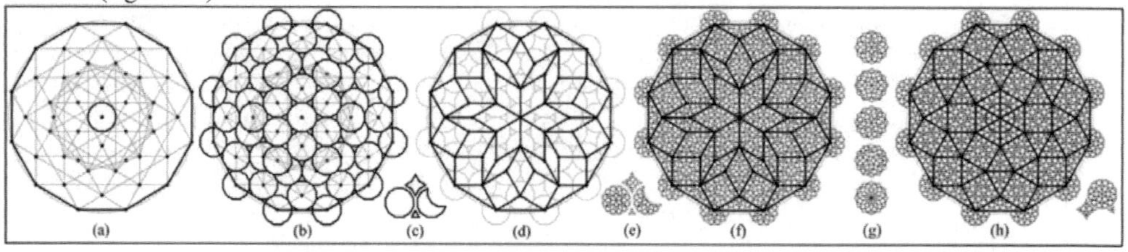

Figure 3. The construction process of the first hierarchy of a new dodecagonal quasiperiodic pattern.

To construct the second hierarchy of the dodecagonal pattern, a new generation of the underlying structural network is constructed, in which, the final constructed pattern of the first hierarchy (figure 3f) acts as a 'seed unit' for the second hierarchy. By mapping the locations of the second generation 'seed units' to the specific intersection points of the structural network, a distribution map of the generalized 'seed polygons' is created (figure 4a). In keeping with a consistent process, three connecting units are needed to fill-in the remaining gaps between the main seed units, all of which, are generated from the main 'seed unit' (figure 4b). The final dodecagonal pattern of the second hierarchy is shown in figure 4c; exhibiting the hierarchical self-similarity attributes.

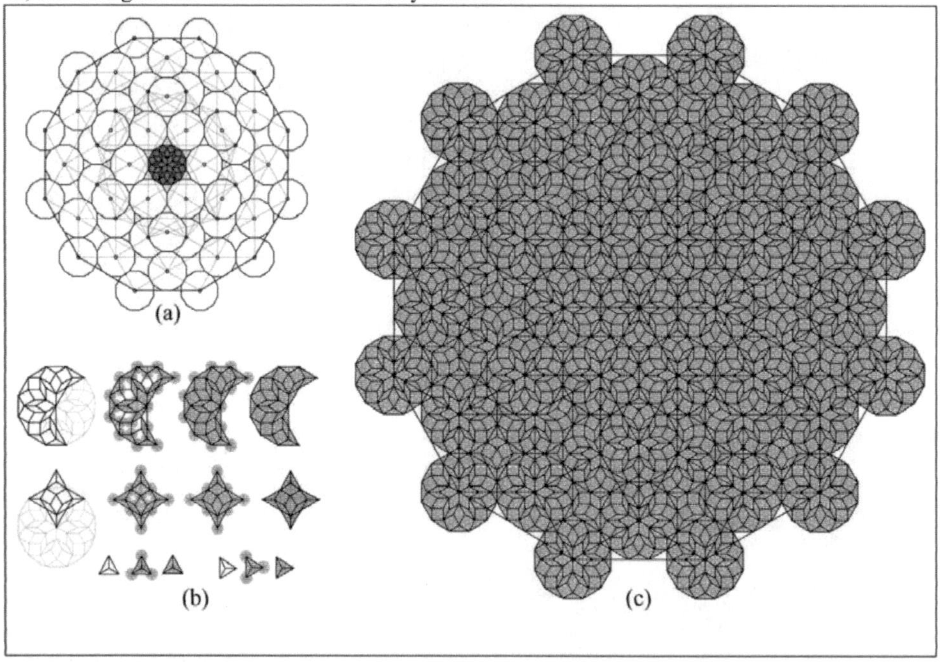

Figure 4. The construction process of the second hierarchy of the dodecagonal pattern.

It is worth noting that this hierarchical quality is a signature characteristic of the natural structure observed in dodecagonal quasicrystals [10]. One of the first dodecagonal quasicrystals ($Ta_{1.6}Te$) was synthesized in 1998 [11, 12]. Figure 5 demonstrates the use of the proposed hierarchical method to describe Conrad et al.'s [11] proposed model for the structure of dodecagonal quasicrystals of $Ta_{1.6}Te$. A framework of nested dodecagrams (figure 5a) forms the structural network for constructing the first hierarchy. The sequence of nested dodecagrams grows based on the square irrational factor $\sqrt{2+\sqrt{3}}$, in which, each larger dodecagram is rotated by 15° degrees. Figure 5c shows the distribution of the seed polygons by this network. The connecting formations, in this case, are derived from the overlapping of the seed polygons (figure 5b). Similarly, the connecting formations of the second hierarchy (figure 5d) are also derived from the main seed cartwheel pattern in Figure 5c. A strong geometrical correlation exists between the structural polygonal map of the second hierarchy and the electron diffractograms of $Ta_{1.6}Te$ quasicrystals [11] (figure 5f). Figure 5g shows that the structural polygonal map of the third hierarchy works in perfect concert with the results of an algorithm for the generation of a self-similar dodecagonal dot pattern produced by Conrad et al. [11].

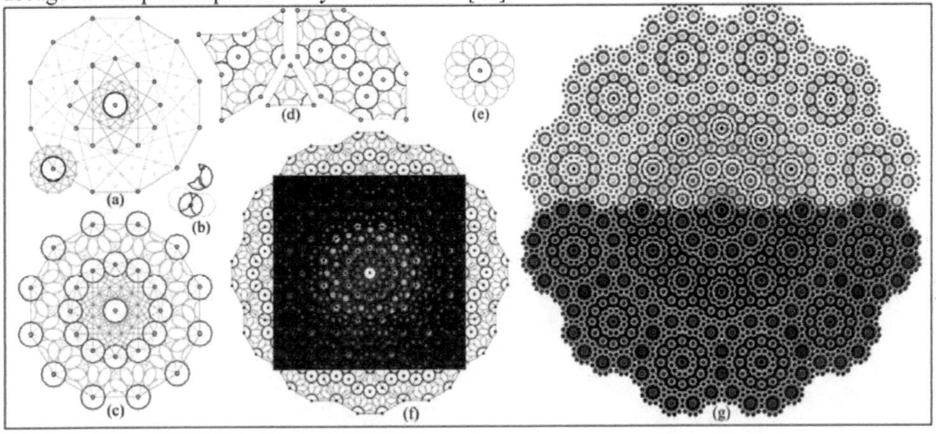

Figure 5. The use of the proposed hierarchical model to describe the structure of dodecagonal quasicrystals of $Ta_{1.6}Te$. The diffractograms in (f) and the algorithm result in (g) are taken from Conrad et al. [11], reproduced with permission from John Wiley and Sons.

4. Conclusion

Derived from the traditional principles of Islamic geometry, this paper presents a global multi-level structural model that is able to describe the underlying structure of dodecagonal patterns in Islamic Architecture. Moreover, this method can be used as a general guiding principle for constructing a wide variety of dodecagonal quasiperiodic patterns without the need for complicated mathematics; providing an easy tool for scientists, mathematicians, teachers, designers and artists, to generate and study a wide range of dodecagonal quasiperiodic patterns.

References

[1] Makovicky E 1992 *Fivefold Symmetry,* ed I Hargittai (Singapore: World Scientific) p 67-86.
[2] Makovicky E, Pérez F R and Hach-Alí P F 1998 *Bol. Soc. Esp. Mineral.* **21** 107–27.
[3] Lu P and Steinhardt P 2007 *Sci.* **315** 1106-10.
[4] Al Ajlouni R 2012 *Acta Crystallogr.* A. **68** 235.
[5] Makovicky E and Hach-Ali P F 1996 *Bol. Soc. Esp. Mineral.* **19** 1–26.
[6] Al Ajlouni R 2013 *Aperiodic Crystals*, ed S Schmid, R L Withers and R Lifshitz (Amsterdam: Springer) p 49.
[7] Makovicky E and Makovicky N 2011 *J. Appl. Cryst.* **44** 569.
[8] Cromwell P R 2015 *Math. Intell.* **37** 30.
[9] Al Ajlouni R 2011 *Philos. Mag.* **91** 2728.
[10] Ishimasa T 2011 *Isr. J. Chem.* **51** 1216.
[11] Conrad M, Krumeich F and Harbrecht B 1998 *Angew. Chem. Int. Ed.* **37** 1383.
[12] Krumeich F, Conrad M, Nissen H U and Harbrecht B 1998 *Philos. Mag. Lett.* **78** 357.

IOP Conf. Series: Journal of Physics: Conf. Series **809** (2017) 012029 doi:10.1088/1742-6596/809/1/012029

Color groups of colorings of N-planar modules

Manuel Joseph C. Loquias, Lilibeth D. Valdez and Ma. Lailani B. Walo

Institute of Mathematics and Natural Sciences Research Institute, University of the Philippines Diliman, 1101 Quezon City, Philippines

E-mail: {mjcloquias, ldicuangco, walo}@math.upd.edu.ph

Abstract. A submodule of a $\mathbb{Z}$-module determines a coloring of the module where each coset of the submodule is associated to a unique color. Given a submodule coloring of a $\mathbb{Z}$-module, the group formed by the symmetries of the module that induces a permutation of colors is referred to as the color group of the coloring. In this contribution, a method to solve for the color groups of colorings of N-planar modules where $N = 4$ and $N = 6$ are given. Examples of colorings of rectangular lattices and of the vertices of the Ammann-Beenker tiling are given to exhibit how these methods may be extended to the general case.

1. Introduction

Colorings and color symmetries of crystals and quasicrystals can be considered a classical topic in crystallography (see [1, 2] and references therein). Most of the literature on colorings of lattices or $\mathbb{Z}$-modules put emphasis on Bravais colorings (*cf.* [3, 4, 5, 6]). Nevertheless, studies have been made on general colorings of lattices or $\mathbb{Z}$-modules such as in [7, 8, 9]. In this contribution, we give some results on how to identify the color groups of sublattice colorings of lattices or $\mathbb{Z}$-modules, and give examples to illustrate them. In particular, we give nonperfect non-Bravais colorings of the rectangular lattice and the vertices of the Ammann-Beenker tiling.

By a *$\mathbb{Z}$-module $\mathcal{M}$ of rank r and dimension d*, we mean a subset of $\mathbb{R}^d$ that is spanned by r vectors in $\mathbb{R}^d$ that are linearly independent over $\mathbb{Z}$ but span $\mathbb{R}^d$. As a group, $\mathcal{M}$ is isomorphic to the free abelian group of rank r. A $\mathbb{Z}$-module Γ of rank $r = d$ is called a *lattice*. At the outset, we shall talk primarily of lattices in dimension d, but results do hold for $\mathbb{Z}$-modules.

Let Γ_1 be a lattice in $\mathbb{R}^d$. A *coloring* of Γ_1 by m colors is an onto mapping $c : \Gamma_1 \to C$, where C is the set of m colors used in the coloring. Of particular interest are colorings of Γ_1 by m colors wherein two points of Γ_1 are assigned the same color if and only if they belong to the same coset of a sublattice Γ_2 of index m in Γ_1. Such a coloring shall be referred to as a *coloring of Γ_1 determined by its sublattice Γ_2*. Here, the set of colors C can be identified with the quotient group Γ_1/Γ_2 so that the color mapping c is simply the canonical projection of Γ_1 onto Γ_1/Γ_2 with kernel Γ_2. Hence, we may take $C = \{c_0 = \mathbf{0}, c_1, \ldots, c_{m-1}\}$ to be a complete set of coset representatives of Γ_2 in Γ_1, and say that the coset $c_j + \Gamma_2$ has color c_j.

Denote by G the symmetry group of Γ_1 and fix a sublattice coloring c of Γ_1. The *point group* of G, $P(G)$, consists of all symmetries in G that fix the origin. A symmetry in G is called a *color symmetry* of the coloring if all and only those points having the same color are mapped by the symmetry to a fixed color. The set H of all color symmetries of the coloring, that is, $H = \{h \in G : \exists \sigma_h \in S_C \text{ such that } \forall \ell \in \Gamma_1, c(h(\ell)) = \sigma_h(c(\ell))\}$, where S_C is the group of permutations on the set of colors C, forms a group called the *color group* or *color symmetry group* of the coloring. The mapping $P : H \to S_C$ with $h \mapsto \sigma_h$ defines a group homomorphism, and thus the group H acts on C. A coloring where $G = H$ is referred to as a *perfect coloring*.

IOP Conf. Series: Journal of Physics: Conf. Series **809** (2017) 012029 doi:10.1088/1742-6596/809/1/012029

The color groups of sublattice colorings of the square and hexagonal lattices were computed in [9]. They were obtained by representing sublattices of the square and hexagonal lattices as upper triangular matrices and expressing elements of the point groups of the lattices as matrices relative to the standard basis of $\mathbb{R}^2$. On the other hand, colorings of the square and hexagonal lattices determined by their square and hexagonal sublattices, respectively, (also referred to as *Bravais colorings*) were considered in [6]. In particular, the color groups were computed using number-theoretic properties of the ring of Gaussian and Eisenstein integers.

2. Some general results

Consider a coloring c of the lattice Γ_1 in $\mathbb{R}^d$ determined by a sublattice Γ_2 of index m in Γ_1. The next theorem, which belongs to the folklore and was proven for specific cases (*cf.* [9, Theorems 1 and 2] and [6, Theorem 2]), gives us a method to compute for the color group H of c.

Theorem 1. *Let G be the symmetry group of Γ_1, where $P(G)$ and $T(G)$ denote the point group and translation subgroup of G, respectively. Then $T(G)$ is a subgroup of the color group H. In addition, an element $g \in P(G)$ is a color symmetry of c if and only if g leaves Γ_2 invariant, that is, $g\Gamma_2 = \Gamma_2$.*

Proof. Let $g \in T(G)$ be the translation by $\ell \in \Gamma_1$. For any $c_i + \Gamma_2 \in \Gamma_1/\Gamma_2$, $g(c_i + \Gamma_2) = (\ell + c_i) + \Gamma_2 = c_j + \Gamma_2$ for some $0 \le j \le m-1$. Hence, g sends color c_i to color c_j, and $g \in H$.

Suppose $g \in P(G)$ is a color symmetry of c. Then g sends color c_0 to some color c_i, $0 \le i \le m-1$, that is, $g\Gamma_2 = c_i + \Gamma_2$. Since $\mathbf{0} \in \Gamma_2$ and $g(\mathbf{0}) = \mathbf{0}$, then $g\Gamma_2 = \Gamma_2$. In the other direction, suppose $g\Gamma_2 = \Gamma_2$. Then for every coset $c_i + \Gamma_2$, $g(c_i + \Gamma_2) = gc_i + g\Gamma_2 = c_j + \Gamma_2$ for some $0 \le j \le m-1$, since $g \in P(G)$. Thus, $g \in H$. □

It follows from above that every color symmetry of c fixes color c_0. Note however that Theorem 1 does not imply that the color group of a sublattice coloring and the symmetry group of the sublattice are the same.

The following corollary relates color groups of colorings determined by similar sublattices.

Corollary 2. *Let $\Gamma_3 = \alpha R\Gamma_2$ where $\alpha \in \mathbb{R}^+$ and $R \in O(d, \mathbb{R})$. Then $P(H') = R[P(H)]R^{-1}$ where H' is the color group of the coloring of Γ_1 determined by Γ_3.*

Proof. Given $g \in P(G)$, $g\Gamma_3 = \Gamma_3 \Leftrightarrow gR\Gamma_2 = R\Gamma_2 \Leftrightarrow R^{-1}gR\Gamma_2 = \Gamma_2$. The claim now follows from Theorem 1. □

3. Color groups of sublattice colorings of planar lattices

Here, we view planar lattices as discrete subsets of $\mathbb{C}$. In this setting, rotations about the origin by θ in the counterclockwise direction correspond to multiplication by the complex number $e^{i\theta}$. We denote such rotations by $R_{e^{i\theta}}$. Also, the reflection T_r along the real axis is associated to complex conjugation.

Suppose Γ_2 is a sublattice of the planar lattice Γ_1. Then $R_{(-1)}\Gamma_2 = -\Gamma_2 = \Gamma_2$. By applying Theorem 1, we obtain the following result.

Proposition 3. *The $180°$-rotation about the origin is a color symmetry of a sublattice coloring of any planar lattice.*

3.1. Sublattice Colorings of the Square and Hexagonal Lattices

We identify the square lattice Γ_1 with the ring of Gaussian integers $\mathbb{Z}[i]$. The symmetry group G of Γ_1 is of type *p4m* (or $*442$ in orbifold notation) and it is symmorphic, that is $G = P(G) \rtimes T(G)$ where $P(G) = \{R_{\pm 1}, R_{\pm i}\} \rtimes \langle T_r \rangle$ and $T(G)$ is the group of translations of G. Given linearly independent vectors $u, v \in \Gamma_1$, let $\Gamma_2 = \langle u, v \rangle_{\mathbb{Z}}$, that is, the sublattice of Γ_1 generated by u and v. Applying Theorem 1 and Proposition 3 in this setting, it suffices to check whether R_i, T_r, and R_iT_r fix the sublattice Γ_2 in order to identify the color group of the coloring of $\mathbb{Z}[i]$ determined by Γ_2. One obtains the following results.

1. Γ_2 is invariant under R_i if and only if $\mathrm{Im}(u\overline{v})$ divides $N(u)$, $N(v)$, and $\mathrm{Re}(u\overline{v})$.

2. Γ_2 is invariant under T_r if and only if $\mathrm{Im}(u\overline{v})$ divides $\mathrm{Im}(u^2)$, $\mathrm{Im}(v^2)$, and $\mathrm{Im}(uv)$.
3. Γ_2 is invariant under R_iT_r if and only if $\mathrm{Im}(u\overline{v})$ divides $\mathrm{Re}(u^2)$, $\mathrm{Re}(v^2)$, and $\mathrm{Re}(uv)$.

These observations yield the following theorem.

Theorem 4. *Let $\Gamma_1 = \mathbb{Z}[i]$ and $\Gamma_2 = \langle u, v\rangle_{\mathbb{Z}} \subseteq \Gamma_1$. Then the color group of the coloring of Γ_1 determined by Γ_2 is of type*

(a) *p4m if $\mathrm{Im}(u\overline{v})$ divides $N(u)$, $N(v)$, $\mathrm{Re}(u\overline{v})$, u^2, v^2, and uv.*
(b) *p4 if $\mathrm{Im}(u\overline{v})$ divides $N(u)$, $N(v)$, and $\mathrm{Re}(u\overline{v})$, and does not divide any of the following: $\mathrm{Im}(u^2)$, $\mathrm{Im}(v^2)$, and $\mathrm{Im}(uv)$.*
(c) *pmm if $\mathrm{Im}(u\overline{v})$ divides $\mathrm{Im}(u^2)$, $\mathrm{Im}(v^2)$, and $\mathrm{Im}(uv)$, and does not divide any of the following: $N(u)$, $N(v)$, and $\mathrm{Re}(u\overline{v})$.*
(d) *cmm if $\mathrm{Im}(u\overline{v})$ divides $\mathrm{Re}(u^2)$, $\mathrm{Re}(v^2)$, and $\mathrm{Re}(uv)$, and does not divide any of the following: $N(u)$, $N(v)$, and $\mathrm{Re}(u\overline{v})$.*
(e) *p2 if none of the above criteria is satisfied.*

If we write $\Gamma_2 = \langle a, b + ci\rangle_{\mathbb{Z}}$, where $a, b, c \in \mathbb{Z}$, then one readily obtains the results in [9] by applying Theorem 4. Because of Corollary 2, we may even assume that a, b, and c are coprime.

Analogous results also hold for colorings of the hexagonal lattice. We associate the hexagonal lattice to the ring of Eisenstein integers $\mathbb{Z}[\xi]$, where $\xi = e^{2\pi i/3}$. For the following result, given $u = a + b\xi \in \mathbb{Z}[\xi]$, we use the notation $\mathcal{R}e(u) := a$ and $\mathcal{I}m(u) := b$.

Theorem 5. *Let $\Gamma_1 = \mathbb{Z}[\xi]$ and $\Gamma_2 = \langle u, v\rangle_{\mathbb{Z}} \subseteq \Gamma_1$. Then the color group of the coloring of Γ_1 determined by Γ_2 is of type*

(a) *p6m if $\mathcal{I}m(u\overline{v})$ divides $N(u)$, $N(v)$, $\mathcal{R}e(u\overline{v})$, u^2, v^2, and uv.*
(b) *p6 if $\mathcal{I}m(u\overline{v})$ divides $N(u)$, $N(v)$, and $\mathcal{R}e(u\overline{v})$, and does not divide any of the following: $\mathcal{I}m(u^2)$, $\mathcal{I}m(v^2)$, and $\mathcal{I}m(uv)$.*
(c) *cmm if $\mathcal{I}m(u\overline{v})$ divides $\mathcal{I}m(u^2)$, $\mathcal{I}m(v^2)$, and $\mathcal{I}m(uv)$, and does not divide any of the following: $N(u)$, $N(v)$, and $\mathcal{R}e(u\overline{v})$.*
(d) *cmm if $\mathcal{I}m(u\overline{v})$ divides $\mathcal{R}e(u^2)$, $\mathcal{R}e(v^2)$, and $\mathcal{R}e(uv)$, and does not divide any of the following: $N(u)$, $N(v)$, and $\mathcal{R}e(u\overline{v})$.*
(e) *cmm if $\mathcal{I}m(u\overline{v})$ divides $\mathcal{R}e(u^2) - \mathcal{I}m(u^2)$, $\mathcal{R}e(v^2) - \mathcal{I}m(v^2)$, and $\mathcal{R}e(uv) - \mathcal{I}m(uv)$, and does not divide any of the following: $N(u)$, $N(v)$, and $\mathcal{R}e(u\overline{v})$.*
(f) *p2 if none of the above criteria is satisfied.*

3.2. Example of colorings of a rectangular lattice

Let Γ_1 be the rectangular lattice given by $\Gamma_1 = \langle 1, 2i\rangle_{\mathbb{Z}}$. For Figure 1(a), $\Gamma_2 = \langle 3, 2i\rangle_{\mathbb{Z}}$. By using the above results to check for the invariance of Γ_2 under T_r, the color group of the coloring of Γ_1 determined by Γ_2 is of type *pmm*. For Figure 1(b), $\Gamma_3 = \langle 4, 1 + 2i\rangle_{\mathbb{Z}}$. Now, by applying the results above, the color group of the coloring of Γ_1 determined by Γ_3 is of type *p2*.

Figure 1. Colorings of $\Gamma_1 = \langle 1, 2i\rangle_{\mathbb{Z}}$ determined by (a) $\Gamma_2 = \langle 3, 2i\rangle_{\mathbb{Z}}$ and (b) $\Gamma_3 = \langle 4, 1 + 2i\rangle_{\mathbb{Z}}$.

IOP Conf. Series: Journal of Physics: Conf. Series **809** (2017) 012029 doi:10.1088/1742-6596/809/1/012029

4. Nonperfect colorings of the Ammann-Beenker tiling

The techniques that we have developed in the previous sections may also be used to obtain submodule colorings of $\mathbb{Z}$-modules. We illustrate this by considering the planar $\mathbb{Z}$-module $\mathcal{M}_1 = \mathbb{Z}[\xi]$ of rank 4 with basis $\{1, \xi, \xi^2, \xi^3\}$ where $\xi = e^{\pi i/4}$. The symmetry group of $\mathbb{Z}[\xi]$ is $G = \langle a, b \rangle \cong D_8$, where a and b correspond to the 45°-rotation in the counterclockwise direction about the origin and reflection along the real axis, respectively. These submodule colorings induce colorings of the vertices of the eight-fold symmetric Ammann-Beenker tiling that are not necessarily perfect.

Given a submodule $\mathcal{M}_2$ of $\mathcal{M}_1$, we identify the color group of the coloring determined by $\mathcal{M}_2$ by checking the invariance of $\mathcal{M}_2$ under R_ξ, R_{ξ^2}, R_{ξ^3}, and T_r. Note that $\mathcal{M}_2$ is always fixed by the 180°-rotation about the origin.

As an illustration, to obtain a coloring of the Ammann-Beenker tiling with color group isomorphic to D_4, we choose the basis elements of $\mathcal{M}_2$ such that $\mathcal{M}_2$ is invariant under R_{ξ^2} and T_r but not under R_ξ. For example, if $\mathcal{M}_2 = \langle 1, 2\xi, \xi^2, 2\xi^3 \rangle$ then it can be verified that $\mathcal{M}_2$ satisfies the given conditions. Thus, the color group of the coloring induced by $\mathcal{M}_2$ is $\langle a^2, b \rangle \cong D_4$ (see Figure 2(a)). The submodule $\mathcal{M}_3 = \langle 1, 2\xi, 2\xi^2, \xi^3 \rangle$ is invariant only under the rotation by 180°. The color group of the coloring induced by $\mathcal{M}_3$ is $\langle a^4 \rangle \cong \mathbb{Z}_2$ (see Figure 2(b)).

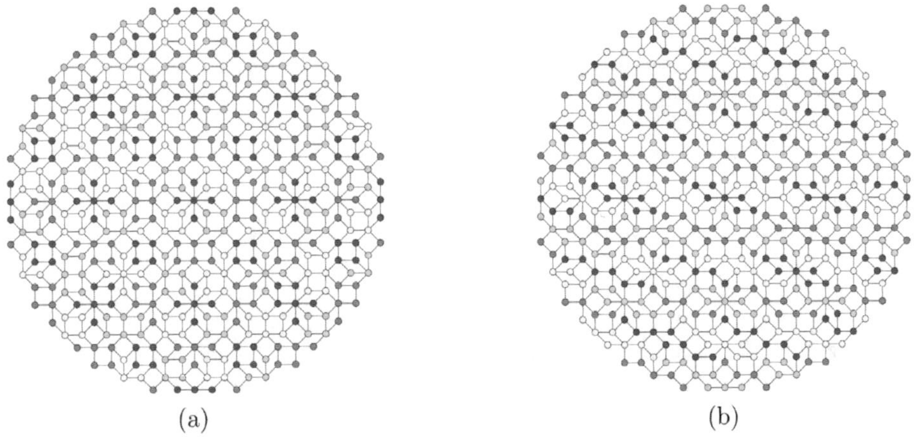

(a) (b)

Figure 2. Colorings of the Ammann-Beenker tiling induced by (a) $\mathcal{M}_2 = \langle 1, 2\xi, \xi^2, 2\xi^3 \rangle$ and (b) $\mathcal{M}_3 = \langle 1, 2\xi, 2\xi^2, \xi^3 \rangle$

Acknowledgments

The authors are grateful for the financial support by the Natural Sciences Research Institute (NSRI) of the University of the Philippines Diliman under Project Code: MAT-12-1-01.

References

[1] Schwarzenberger R L E 1984 Colour symmetry *Bull. Lond. Math. Soc.* **16** 209-40
[2] Lifshitz R 1997 Theory of color symmetry for periodic and quasiperiodic crystals *Rev. Mod. Phys.* **69** 1181-1218
[3] Baake M 1997 Combinatorial aspects of colour symmetries *J. Phys.* A **30** 2687-98
[4] Baake M, Grimm U and Scheffer M 2002 Colourings of planar quasicrystals *J. Alloys Compounds* **342** 195-97
[5] Baake M and Grimm U 2004 Bravais colourings of planar modules with N-fold symmetry *Z. Krist.* **219** 72-80
[6] Bugarin E P, De Las Peñas M L A and Frettlöh D 2013 Perfect colourings of cyclotomic integers *Geom. Dedicata* **162** 271-82
[7] Harker D 1978 Colored lattices *Proc. Natl. Acad. Sci. U.S.A.* **75** 5264-67
[8] Moody R V and Patera J 1994 Colourings of quasicrystals *Can. J. Phys.* **72** 442-52
[9] De Las Peñas M L A and Felix R 2007 Color groups associated with square and hexagonal lattices *Z. Krist.* **222** 505-12

Electron Localization and Conduction Mechanism of Icosahedral Quasicrystal Al-Pd-Ru

Y.Sekiyama and I.Kanazawa

Department of Physics, Tokyo Gakugei University, Tokyo 184-8501, Japan

Kanazawa@u-gakugei.ac.jp

Abstract. We have argued the temperature dependence of conductivities of the icosahedral Al-Pd-Ru quasicrystals from the standpoint of the quasicrystal-like system with sp-d hybridization, using the thermal Green's function technique.

1. Introduction

A variety of Al-based quasicrystals shows unusual electronic properties which seem to be different from those of metals and semiconductors.

One of the most anomalous properties of Al-based quasicrystals is a very low electrical conductivity close to the metal-insulator transition [1,2]. However, there is no evidence for a gap as in semiconductors, and the structure is highly ordered, although it is nonperiodic. These are unlike Anderson insulators, where disorder is the reason for a localization of electronic wave function.

It has been suggested that the anomalous conductivity of Al-based quasicrystals is due to the combination of a Hume-Rothery pseudogap and the presence of localized states at the Fermi level [3].

It has been proposed that the unusual transport properties of Al-based quasicrystals might be much related to hopping processes between wave functions localized inside atomic clusters [4,5].

Especially Trambly de Laissardiere et al. [5] argued the creation of virtual bond states due to an icosahedral cluster of transition metal (TM) atoms and an icosahedral cluster of the TM clusters. In addition, in views of the ab initio calculations [6,7], Tamura et al. [8] have suggested strongly the creation of the virtual bond states or confinement of electrons, in the icosahedral clusters of TM atoms.

Kanazawa and coworkers [9-16] have considered the transport property in the randomly distributed system of the cluster composed of correlated N number of configurations such as the prolate and oblate rhombohedra, in which the nearest distance between each configuration is $\sim 2\pi/2k_F$ (the quasicrystal-like state). The quasicrystal-like state is regarded as the system composed of the Gaussian correlated distribution of the icosahedral cluster such as the Bergman type and the Mackay type, which include $2k_F$-phase shift scattering.

Taking into account the mean free path 15~20Å of electrons in quasicrystals, it looks like that the anomalous transport properties of quasicrystals are not directly related to quasicrystal structure in the longer range than the mean free path 15~20Å. Thus the transport property in the quasicrystal-like system might be closely analogous to one in the quasicrystal.

Tamura et al. [17] have measured the temperature dependence of the resistivity of the icosahedral Al-Pd-Ru quasicrystals, and 1/0 and 2/1 cubic approximant phases.

In this study, we have analyzed the temperature dependence of the conductivities of the icosahedral Al-Pd-Ru quasicrystals from the standpoint of the quasicrystal-like system, using the thermal Green's function technique, and taking into account the sp-d hybridization.

2. A model system and electrical conductivities in Al-Pd-Ru quasicrystals

In the quasicrystal-like system, we shall consider the effect of the $2k_F$-phase shift scattering by the randomly distributed icosahedral cluster composed of correlated N number of configuration such as the prolate and oblate rhombohedra, in which the nearest distance between each configuration is $\sim 2\pi/2k_F$, in the Feynman graph with thermal Green's function technique.

It has been proposed that the sp-d hybridization plays an important role in the transport property of quasicrystals [18,19].

The $2k_F$-phase shift scattering induces a strong density wave of sp electrons with wavelength $\sim 2\pi/2k_F$.

When the high density region of the standing wave of sp electrons is located on the transition metal atoms, the sp electrons hybrize more strongly the d orbitals of transition metal atoms. Thus $2k_F$-phase shift scattering and the sp-d hybridization are much correlated to each other. In this case, the value γ, which is defined as $1/[2\pi\rho\tau]$, becomes large. Note that $\gamma \propto n_i N|V_{d,sp}|^2$. Here ρ is the density of state per spin at the Fermi energy and τ is the life time of the sp electron wave due to the sp-d hybridization et al. The matrix element $V_{d,sp}$ represents the sp-d hybridization, and n_i is the density of the icosahedral cluster, which is composed of N number of configurations such as the prolate and oblate rhombohedra, connection by distance $\sim 2\pi/2k_F$. This icosahedral cluster might be identified with one such as Bergman type and the Mackay type [20-22]. N depends on the kind of materials. When many defects are introduced in the system, the icosahedral cluster, which is composed of correlated N number of configurations connected with the distance $\sim 2\pi/2k_F$, is broken. As a result, the value of γ will decrease remarkable and the resistivity of the system decreases.

Performing an average over the position of the icosahedral cluster, we assume the potential of the icosahedral cluster as a random quantity having a Gaussian δ-correlated distribution,

$$\langle V(r)V(r')\rangle = \frac{n_i N|V_{d,sp}|^2}{2\pi\rho}\delta(r-r')$$

Where $V(r)$ is the effective potential of the clusters, and ρ is the state density per spin at the Fermi energy. The propagator $\Gamma(q,\omega_l)$ of the $2k_F$-phase shift scattering with the sp-d hybridization can be introduced as

$$\Gamma(q,\omega_l) \sim \frac{\gamma}{1-\gamma\Pi(q,\omega_l)}, \tag{1}$$

where, ω_l is $2\pi lT$. l is an integer. $\gamma = \frac{n_i N|V_{d,sp}|^2}{V_s}$. V_s is volume of the system.

The thermal Green's function $D(k,\varepsilon_n)$ ($\varepsilon_n = (2n+1)\pi T, T$ is the temperature) is given by

$$D(k,\varepsilon_n) = \left[i\varepsilon_n - \xi_k + \frac{i}{2\tau}\mathrm{sgn}\varepsilon_n\right]^{-1}, \tag{2}$$

$$\frac{1}{\tau} \sim 2\pi\rho\gamma$$

where $\xi_k = k^2/2m - E_F$.

$\prod(q,\omega_l)$ is represented as

$$\prod(q,\omega_l) = \sum_k D(k+q-2k_F,\varepsilon_n+\omega_l)D(k,\varepsilon_n)$$

$$\sim\rho\int_{-\infty}^{\infty} d\xi \int d\Omega\left[i\varepsilon_n + i\omega_l - \xi - vq + \frac{i}{2\tau}\mathrm{sgn}(\varepsilon_n+\omega_l)\right]^{-1}$$

$$\times \left[i\varepsilon_n - \xi + \frac{i}{2\tau} \mathrm{sgn}(\varepsilon_n) \right]^{-1} , \quad (3)$$

where $\xi_{-k} = \xi_k$ and $v \equiv k/m$, and k_F is the vector of the Fermi momentum.

The integration over $d\Omega$ is over angle of v. If $\varepsilon_n(\varepsilon_n + \omega_l) < 0$, $\Pi(q, \omega_l)$ is evaluated for small q and ω_l as

$$\Pi(q, \omega_l) \sim 2\pi\rho\tau[1 - Dq^2\tau - |\omega_l|\tau] , \quad (4)$$

where $D = v_F^2\tau/d$ is the diffusion constant. v_F and d are Fermi velocity and dimensions of system, respectively. Thus, $\Gamma(q, \omega_l)$ is represented as

$$\Gamma(q, \omega_l) = \frac{1}{2\pi\rho\tau^2[Dq^2 + |\omega_l|]} . \quad (5)$$

By using $\Gamma(q, \omega_l)$, the replacement $\omega_l \to i\omega$, and the Green's functions $D_\pm(k)$ we can estimate the conductivity approximately in $d = 3$,

$$\Delta\sigma(\omega) = \frac{e^2}{\pi} \sum_k \sum_{k'} (\hbar/m)^2 k_x k'_x D_+(k) D_-(k) D_+(k') D_-(k') \Gamma(k + k' = q, -i\omega) . \quad (6)$$

$$\propto \frac{e^2}{\hbar^2} \sqrt{\frac{\Gamma}{D}} T. \quad (7)$$

This shows T-liner effect for the conductivity.

We shall analyze the conductivity of Al-Pd-Ru quasicrystals of Tamura et al. [17] in order to discuss the effect of the electron localization in the quasicrystal-like system.

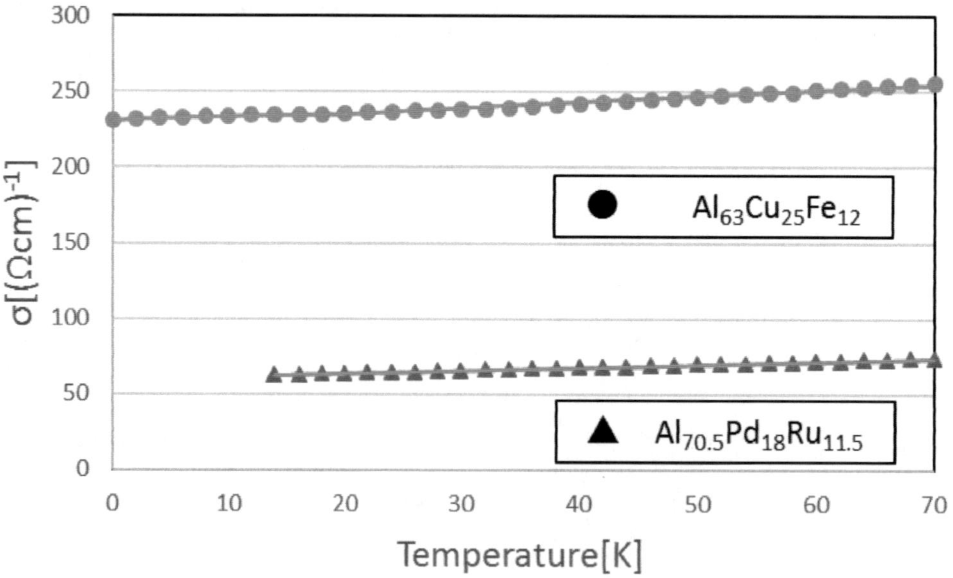

Figure 1. Solid triangles represent data of conductivities of $Al_{70.5}Pd_{18}Ru_{11.5}$ quasicrystal [17]. Solid circles represent data of conductivities of $Al_{63}Cu_{25}Fe_{12}$ quasicrystal [23]. Solid lines show fitted ones.

Figure 1 shows the conductivity for the $Al_{70.5}Pd_{18}Ru_{11.5}$ quasicrystal.

Each data point is represented by the solid triangle [17]. The solid line shows the fit. For comparison, experimental data and fitting of the $Al_{63}Cu_{25}Fe_{12}$ quasicrystal [15,23] are also shown in Fig.1.

As the inelastic scattering length $L_{int}(T)$ decreases when temperature increases, samples can leave from the metallic regime at low T into the critical regime at enough high T. Taking into account the analysis by scaling of the conductivity in quasicrystals [24], we think that the Al-Pd-Ru quasicrystal in Fig.1 is on the metallic side, where the condition, $L_{int}(T)$ > the electron mean free path is satisfied below ~70K. That is, it is inadequate that the temperature curves for the conductivity σ of the Al-Pd-Ru quasicrystal in the temperature region above ~70K are analyzed by the weak localization theory or the present theoretical formula. A solid line for Al-Pd-Ru quasicrystal in Fig.1 shows the temperature curve for σ described by a power law $\sigma(T) = a + bT$. This relation is consistent with T-liner effect in eq.(7) for the conductivity, which is derived theoretically from the quasicrystal-like system. A solid line for Al-Cu-Fe quasicrystal was fitted by a power law $\sigma(T) = c + dT^{3/4}$ from 20 to 70K [15]. $T^{3/4}$ dependence is introduced from higher-order perturbation effect of conducting mechanism in the quasicrystal-like system.

In addition, we stress strongly that the present conducting mechanism can explain evidently the reduction of the resistivity due to introducing many defects in quasicrystals.

3. Conclusion

The T-liner power law fits well conductivity of Al-Pd-Ru quasicrystal below 70K.

It is suggested strongly that the anomalous transport properties are much related to localization effect in the quasicrystal-like system.

.

References

[1] Poon S J, 1992 *Adv. Phys.* **41**, 303
[2] Klein T et al. 1991 *Phys. Rev. Lett.* **66**, 2907
[3] Berger C, in LECTURES ON QUASICRYSTALS (Eds. D.Gratias Les Editions de physique Les Ulis, 1994) p 463
[4] Mayou D et al. 1993 *Phys. Rev. Lett.* **70**, 3915
[5] Trambly G de Laissardiere and Mayou D, 1997 *Phys. Rev.* **B55**, 2890
[6] Fujiwara T and Yokokawa T, 1991 *Phys. Rev. Lett.* **66**, 333
[7] Hafner J and Crajci M, 1992 *Phys. Rev. Lett.* **68**, 2321
[8] Tamura R et al. 2001 *Phys. Rev. Lett.* **68**, 3104
[9] Kanazawa I, 2000 *Physica B* **284-288**, 1161
[10] Kanazawa I, 2002 *J.Alloys Compounds* **342**, 381
[11] Kanazawa I, 2003 *Physica B* **328**, 111
[12] Kanazawa I, 2003 *Mod. Phys. Lett.* **B 17** 841
[13] Kanazawa I, 2003 *Physica E* **18**, 306
[14] Akiyama K, Takagiwa Y, Kanazawa I, 2006 *Philo. Mag.* **86**, 747
[15] Kanazawa I, Kitahata H, Sekiyama Y, 2008 *Z.Kristallog.* **223**, 813
[16] Yamada K et al. 2012 *Physica Procedia* **35**, 51
[17] Tamura R et al. 1999 *J.Phys:Condens. Matter* **11**, 10343
[18] Trambly G de Laissardiere et al. 1993 *Europhys. Lett.* **21**, 319
[19] Takeuchi T et al. 1998 *Phys. Rev.* **B58**, 11345
[20] Janot C and Boissien M, 1994 *Phys. Rev. Lett.* **72**, 1674
[21] Janot C, 1996 *Phys. Rev.* **B58**, 181
[22] Katz A and Gratias D, 1994 *J.Non-Cryst. Solids* **153-154,** 184
[23] Klein T et al. 1992 *Phys. Rev.* **B45**, 2046
[24] Delahaye J and Berger C, 2001 *Phys. Rev.* **B64**, 094203